MTI Publication No. T-2

Hydrodynamic Modeling of Corrosion of Carbon Steels and Cast Irons in Sulfuric Acid

by

M. Rahmani and J.E. Strutt

School of Industrial Science
Cranfield Institute of Technology
Cranfield, Bedford MK43 0AL
United Kingdom

Published for The Materials Technology Institute
of the Chemical Process Industries, Inc.
by the National Association of Corrosion Engineers

Library of Congress Catalog Card Number: 92-80991

ISBN: 1-877914-40-1
Printed in the United States of America

This document was prepared under the sponsorship of Materials Technology Institute of the Chemical Process Industries, Inc. (MTI) and is approved for release. All data and information contained in this document are believed to be reliable; however, no warranty of any kind, express or implied, with respect to the data, analyses, recommendations or any other information provided is made by MTI or the author of this document and the use of any part of this document is at the user's sole risk. MTI, the author, or any person acting on their behalf, assume no liability and expressly disclaim liability, including without limitation liability for negligence, resulting from the use or publication of the information contained in this document or warrant that such use or publication will be free from privately owned rights. Request for republication rights should be addressed to the Executive Director of MTI.

Published for The Materials Technology Institute
of the Chemical Process Industries, Inc.
By The National Association of Corrosion Engineers
P.O. Box 218340
Houston, TX 77218-8340

Information on availability of this document can be obtained from NACE.

ABOUT MTI

The Materials Technology Institute of the Chemical Process Industries, Inc. (MTI) is a unique, cooperative research organization representing private industry. Its goal is to conduct generic, nonproprietary studies of a practical nature on the deterioration of materials and equipment used in the process industries.

Through membership in MTI, companies can solve nonproprietary problems of major concern to the process industries, leverage research dollars by participating in the direction and results of MTI studies, capitalize on the expertise of member company representatives, and learn about MTI's accomplishments in a timely fashion.

More information about membership in MTI can be obtained from the Executive Director. MTI members receive copies of MTI publications as part of their membership.

The Materials Technology Institute
of the Chemical Process Industries, Inc.
1215 Fern Ridge Parkway, Suite 116
St. Louis, MO 63141-4401
314/576-7712

CONTENTS

LIST OF TABLES

LIST OF FIGURES

NOTATIONS

A	Conversion factor between mass transfer rate (moles/cm_2/s) and corrosion rate (mm/yr^{-1}) : $2.24 \cdot 10^9$
a,b,c	Constants
c_b	Ferrous ion concentration in bulk acid (mols/cm^3)
c_s	Ferrous ion solubility in sulfuric acid, electrode surface concentration (mols/cm^3)
D	Diffusion coefficient of ferrous ion in sulfuric acid (cm^2/s)
Df	Diffusion coefficient of ferrous ion in surface film (cm^2/s)
D_{eff}	Effective diffusion coefficient of ferrous ion in surface film (cm^2/s)
D_{TURB}	Turbulent eddies diffusion coefficient (cm^2/s)
d	Diameter of inner rotating cylinder: 3.1 cm
E	Corrosion potential (mV)
e	Surface film thickness (cm)
F	Farady's constant: 96,500 Coulombs/equivalent
f	Friction factor (dimensionless)
f( )	Function of ...
g( )	Function of ...
h	Height of knurl on rough cylinder
I_{LCALC}, I_{LMEAS}	Calculated and measured limiting corrosion current (mA)
i_L	Limiting corrosion current density (mA/cm^2)
j	Molar flux of reacting species (moles/cm^2/s)

ive.

K	[illegible]
K	[illegible]
[illegible]	[illegible]
K_{RCE}	Mass transfer factor (cm/s) in the [illegible]
K_{MCR}, K^2	Overall measured mass transfer factor [illegible]
LCDT	Limiting current density technique
L_{MIN}	Minimum entry length for establishment of constant hydrodynamic flow in pipe geometry
M	Atomic weight of iron: 55.85 g
m,n	Integer constants 1,2,3
N	Mass transfer rate [illegible]
Nu	Nusselt number [illegible]
Pr	Prandtl number [illegible]
Re	Reynolds number: $dv\rho/\mu$ (dimensionless)
Re_{crit}	Critical Re for laminar [illegible]
RCE	[illegible]
rpm,S	Rotation per minute
R_f	Film resistance
R_{sol}	Solution resistance
S	Working electrode surface [illegible]
Sc	Schmidt number: [illegible] $\frac{\mu}{\rho D}$
Sh	Sherwood number: $\frac{Kd}{D}$ (dimensionless)
St	Stanton number: [illegible] (dimensionless)

EXECUTIVE SUMMARY

This report presents the results and interpretation of a study to assess the hydrodynamic corrosion test method using the rotating cylinder electrode (RCE) technique. The study was initiated and supported by the Materials Technology Institute of the Chemical Process Industries, Inc. (MTI) and was conducted at Cranfield Institute of Technology between 1984 and 1987. The primary motive behind the study was to assess the use of the RCE technique on an industrially important corrosion system; namely, carbon steels in sulfuric acid.

The interest in using rotating cylinders for studying velocity-dependent corrosion processes stems from the high cost and experimental difficulties of undertaking corrosion measurements on flow loops. While flow loops are more directly relevant to chemical plant pipework, the operationally simpler rotating cylinder, with its much smaller volumes of corrosion media, offers the potential for more cost effective studies of corrosion behavior under well-defined hydrodynamic conditions. For this purpose, an RCE apparatus was built and tested to enable this study to be conducted.

The RCE was designed such that the auxiliary electrode was mounted on the rotating shaft rather than as an outer concentric cylinder. This configuration, which has both practical and economic implications, did not significantly influence the hydrodynamic conditions near the wall of the shaft.

The study has concentrated on industrially relevant systems; namely, steels in concentrated sulfuric acid at 93% and 98.5% strength at temperatures up to 40°C (104°F) and velocities up to 2 ms^{-1}. In order to generate information of practical use to plant corrosion engineers, a widely used steel and a number of cast irons was used in the study:

AISI 1020 (UNS G10200) carbon steel
Ductile iron
Grey cast iron
Mondi A[(1)]
Mondi B[(2)]

Carbon steels and cast irons are extensively used in handling concentrated sulfuric acid under flowing conditions. In the past, the prediction of their corrosion performance in plants has been hampered by the lack of suitable models and data for these systems largely as a result of the complexity of the corrosion process and its dependence on velocity and geometry.

Previous work on these systems has shown that the corrosion mechanism for carbon steel exhibits a significant dependence on mass transfer. This has lead various workers to suggest the use of

---

[(1)]Trade name.
[(2)]Trade name.

dimensionless analysis as a means of relating laboratory-scale experiments to plant-scale corrosion behavior. This is also the approach taken in this study.

The body of the report is in two parts. The Introduction and Chapters One and Two introduce the background theory and critically review prior corrosion research involving mass transfer measurements. Chapters Three, Four, and Five present the experimental work conducted, the results, and an analysis and discussion of the results.

There have been few previous investigations of corrosion involving steels in concentrated sulfuric acid reported in the open literature. However, one investigation of mild steel in 68% sulfuric acid using the RCE had shown that the corrosion rate was proportional to the velocity of the acid raised to the power of 0.7 and that the mass transfer processes could be described by the equation:

$$Sh = 0.079 Re^{0.7} Sc^{0.356}$$

It has also been suggested that this same expression was valid for all acid strengths between 50% and 99%. The present study has confirmed the findings at 68% but has discovered a significant variation of the velocity exponent both with acid strength and temperature.

For carbon steels, the results of this study have shown that convective diffusion of ferrous ion away from a saturated ferrous sulfate film is the rate controlling factor and that the mass transfer process can be represented by the following expressions:

$$Sh = 0.079 Re^{0.78} Sc^{0.356} \text{ at } 27°C\ (104°F) \text{ in } 93\% \text{ and } 98\% \text{ acid}$$
$$Sh = 0.079 Re^{0.81} Sc^{0.356} \text{ at } 40°C\ (81°F) \text{ in } 93\% \text{ and } 98\% \text{ acid}$$

This leads to a significant increase in estimated corrosion rates over those using equations with $Re^{0.7}$ dependence.

The above-mentioned dependence of the velocity exponent on concentration and temperature, as observed using the RCE, may well be occurring in other geometries, pàrticularly the pipe geometry. Unfortunately, time and resources were insufficient to investigate this during the present project, but it is suggested that this study should be checked in future studies of this system.

As mentioned above, the corrosion behavior of the 1020 alloy could be adequately represented by a modified convective diffusion model. By contrast, the cast irons had a tendency to develop films that introduced time dependence in the measured potentials and corrosion rates, resulting in a significant deviation from the mass transfer model to which the 1020 alloy adhered. For cast irons, the deviation at 40°C (104°F) was substantial.

The main consequence of filming is a reduction in corrosion rate below that expected of a mass transfer-controlled reaction and a loss of flow sensitivity. These deviations were found to increase with increased rotational velocity, temperature, and time of exposure. These effects were more pronounced for grey cast iron and Mondi B than for ductile iron and Mondi A. The actual velocity dependence for these alloys was found to depend upon the morphology of the surface film in a complex manner.

A modified model for the corrosion mass transfer equation was employed to analyze and interpret the corrosion behavior of the cast irons. This equation includes an additional mass transfer term to accommodate the diffusional impedance of the films.

At the higher temperature of 40°C (104°F), a mass transfer expression was derived for each of the cast iron alloys as follows:

Grey Cast Iron	$K_{film} = 1.27 \cdot 10^{6} Re^{-2.52} \frac{cm}{s}$
Mondi B	$K_{film} = 8.35 \cdot 10^{3} re^{-1.94} \frac{cm}{s}$
Mondi A	$K_{film} = 8.81 \cdot 10^{2} Re^{-1.63} \frac{cm}{s}$
Ductile Iron	$K_{film} = 8.37 \cdot 10^{5} Re^{-2.39} \frac{cm}{s}$

At 27°C (81°F), grey cast iron showed behavior similar to that observed at 40°C (104°F); namely, a significant reduction in velocity dependence due to filming. It was also found that different electrodes would form films with different mass transfer behaviors. This was particularly so for the grey cast iron. The film mass transfer factor for two different electrodes was calculated to be:

Grey Cast Iron

$$(1)\ K_{film} = 6.29 \cdot 10^{-3} Re^{-0.39} \frac{cm}{s}$$

$$(2)\ K_{film} = 8.90 \cdot 10^{-3} Re^{-0.32} \frac{cm}{s}$$

At 27°C (81°F), Mondi B also exhibited similar behavior to grey cast iron, but in this case, a value for the film mass transfer could not be estimated due to uncharacteristic film growth.

On the other hand, Mondi A and ductile iron did not show a significant loss in velocity dependence and had sufficiently high filming mass transfer factors for the rate limiting step to be mass transfer through the electrolyte rather than the film.

It is concluded that the convective diffusion model modified to accommodate film mass transfer processes can be used to predict the corrosion rate of cast iron alloys, provided that a value for the film mass transfer factor is available at the operating conditions of interest.

ACKNOWLEDGMENTS

The authors would like to thank the Materials Technology Institute of the Chemical Process Industries, Inc. for financial support in conducting this work and Professor P. Hancock for provision of laboratory facilities.

M. Rahmani, would like to thank CERHYD for the leave of absence that was necessary in order to conduct the research and to study for a Ph.D. at Cranfield.

The authors would also like to express their gratitude to F. Parsons for excellent assistance in the laboratory and for having typed the manuscript so accurately.

INTRODUCTION

Sulfuric acid is the most widely used reagent in the chemical industry and far outpaces the second most-used chemical, ammonia. For example, the U.S. Department of Commerce reported the 1962 production of sulfuric acid to be 19,116,828 tons, while production of synthetic ammonia, the closest competitor, was only 5,779,640 tons.[1]

Sulfuric acid is made by the Contact Process. Sulfur is burned to form sulfur dioxide, which is converted to sulfur trioxide by contact with a vanadium catalyst. The sulfur trioxide is then absorbed in sulfuric acid to give various grades of sulfuric acid, such as concentrated oil of vitriol (COV) and rectified oil of vitriol (ROV), 93 to 98.5% sulfuric acid. These acids are consumed in the production of fertilizers, chemicals, petroleum refining, paints, iron and steel, rayon and cellulose fibers, explosives, and a wide variety of other items.

Ordinary carbon steels and cast irons are, on a tonnage basis, the most widely used materials in the production and handling of sulfuric acid at concentrations of over approximately 70% at ambient temperatures and low velocities.[2] Storage tanks, pipelines, tank cars, and shipping drums made of steel are very common for 78%, 93%, 98%, and stronger acids such as oleum (100% and above). The formation of a protective and mildly soluble ferrous sulfate film is responsible for the corrosion resistance observed in this concentration range. A poorer corrosion resistance is obtained at lower acid concentrations because of their higher ferrous sulfate solubilities.[3]

Many corrosion problems have been observed in the industrial use of sulfuric acid. For example, steel is susceptible to hydrogen grooving in the higher concentrations of sulfuric acid. This type of corrosion, which occurs in stagnant pipelines, tank cars, and storage tanks,[4,5] is associated with the rising hydrogen bubbles creating turbulence and resulting in increased mass transfer and corrosion along their path. In a horizontal vessel or pipeline, grooving starts at mid-diameter and forms a deep groove along the top of the vessel or pipeline. The addition of copper to steel reduces this attack.[4,6] Anodic polarization has equally been reported to arrest hydrogen grooving.[4]

Catastrophic failures of storage tanks attributed to inlet erosion have also been reported.[4,7] In these instances, the roof inlet nozzle on 93% acid storage tanks was located close to the tank sidewall. Turbulence, resulting in high mass transfer and corrosion rates, was correctly identified as the cause of failure. To compensate for and reduce this accelerated attack at inlet nozzles, it is common practice to locate roof inlet nozzles as far as possible from the tank sidewall and to install splash plates on the tank floor below the inlet nozzle. It is also recommended that outlet nozzles be extended beyond the tank sidewall and into the bulk acid to prevent accelerated attack of the sidewall. The use of corrosion resistant alloys or heavier wall nozzles is also recommended.[8]

Other problems encountered in the industry include accelerated attack caused by absorption of water from the air.[5,8] Vents on steel storage tanks for strong sulfuric acid are sometimes equipped with special dehydrating devices to prevent this dilution effect.

Preferential weld attack is also a concern.[1,8] The problem can be alleviated by small additions of copper, typically 0.2% Cu, to the steel.

Hydrodynamic corrosion of carbon steel by sulfuric acid is very complex and depends on several parameters; namely, temperature, component geometry, flow velocity, acid concentration, and impurity level, particularly ferrous ion content. Previous work has shown that the corrosion mechanism contains a high dependence on mass transfer.[9-12] This has suggested the introduction of dimensionless analysis developed and used by chemical engineers in the field of mass transfer (and also heat transfer) as a generalizing technique to describe diffusion-controlled corrosion processes.

Dimensionless analysis, in mass transfer studies, is an analytical method devised to introduce all the relevant parameters of the system in a single, simple relation. In as much as the measured variables (fluid velocity, viscosity, density, concentration, etc.) define the system in a specific sense, the dimensionless groups define the general situation allowing general conclusions to be drawn.

Of the dimensionless numbers currently used in mass transfer studies, those of most interest in the corrosion field are the Reynolds number (Re [Re defines the flow regime]), the Schmidt number (Sc [Sc defines the momentum-to-mass-transfer ratio]), and the Sherwood number (Sh [Sh defines the mass transfer rate and is thus directly related to the corrosion rate]). From a detailed analysis of fundamental equations, it is possible to derive mass transfer relationships of the form:

$$Sh = aRe^{b}Sc^{c}$$

where the values of the constants a, b, and c are characteristic of a specific flow system.[1] These can be derived theoretically; however, for practical systems, they are determined empirically.

This type of approach for studying corrosion in flowing environments can be used in the laboratory to gain insight into the corrosion mechanism governing the system of interest provided that it is under mass transfer control. However, it should be recognized that this technique has no validity if the fluid velocity effect is, for example, caused by erosion rather than convective mass transfer. Under these conditions, knowledge of Sh has no relevance with respect to the corrosion rate.

In order to measure corrosion rates for systems that are controlled by mass transport, a laboratory tool is required that simulates the fluid flow in the plant situation. The apparatus must have well-defined hydrodynamics to accurately simulate the corrosion phenomena that occur in the plant environment.

This report is concerned with the implementation of the methods outlined above for the study of the corrosion of carbon steel and cast iron in concentrated sulfuric acid. To achieve this goal, a rotating cylinder apparatus has been designed, built, and used as a laboratory tool to study the corrosion of carbon steel and a number of cast irons in sulfuric acid of 93% and 98% concentration over a range of

---

[1] See "Notations," p. xv.

temperatures and velocities consistent with the use and limitations of the materials in service.

CHAPTER

ONE

BACKGROUND THEORY

In this chapter, the fundamental aspects of fluid flow and transport properties in a fluid are introduced. Also, the different cases of corrosion rate control are reviewed. These will facilitate the definition of the mass transfer factor and of the dimensionless mass transfer relationships. The concept of the limiting current density is also introduced, and its relation to the mass transfer factor, which is the basis of the limiting current density technique (LCDT), is explained.

1.1 FLUID FLOW CHARACTERIZATION

For an accurate study of the influence of flow velocity upon the corrosion rate of fluids in motion, the hydrodynamic conditions must be well-defined. The Reynolds number (Re), a dimensionless number dependent on the fluid velocity or the electrode rotation rate, as the case may be, density and viscosity of the fluid, and a characteristic dimension (i.e., diameter of pipe in pipe flow) is used to define the type of flow. It can be written as:

$$Re = \frac{dv\rho}{\mu} \tag{1.1}$$

At low velocities, i.e., at low Re, a stable or streamlined type of flow or laminar flow is encountered. Assuming the fluids under consideration to be Newtonian and incompressible in nature, the shear stress at any point in a laminar flow is given by:

$$\zeta = \mu \frac{dv}{dy} \tag{1.2}$$

If the velocity is increased, at a critical Reynolds number (Re_{CRIT}), the flow becomes turbulent and an additional mechanism of momentum transfer appears, which is caused by rapid and random fluctuations of velocity about its average value. A common way to express the viscosity in turbulent flow has been to introduce the concept of an eddy viscosity

(ϵ), such that:[13]

$$\mu^{(T)} = \mu + \rho\epsilon \qquad (1.3)$$

Clearly, as turbulent eddies are damped near a wall, ϵ will depend on velocity and distance from the wall.[14] This further implies that the shear stress at the fluid/solid interface:

$$\zeta = \mu^{(T)} \frac{dv}{dy} \qquad (1.4)$$

is greater in turbulent flow.

It has also been suggested that this may control the mechanical stability of films.[15] For many systems, the corrosion rate is controlled by the presence of a film of corrosion products. This may range from the relatively thin films on metals, such as aluminum and magnesium, to massive deposits found on corroding iron pipes.[16,17] The film stability will usually be determined by a combination of electrochemical, thermodynamic (solubility, adsorption), and hydrodynamic conditions. The film structure and bonding to the substrate will also play a role.

In cases of increased turbulence, the higher stress created by the flow at the liquid/solid interface may assume the major role by shear-stripping protective films in some form of enhanced mass transfer or erosion-corrosion mechanism, thus leading to higher corrosion rates than those that could be expected from chemical dissolution alone.[18]

The Re_{CRIT} for the transition between laminar and turbulent flow will vary depending on the geometry, and R_{CRIT} for pipe flow has been experimentally found to be around 2,100, whereas with a rotating disc electrode (RDE), laminar flow can be maintained up to Re values of 10^4 to 10^5.[19]

1.2 EFFECT OF FLOW ON TRANSPORT PROPERTIES IN A FLUID

The interface between a stationary solid (e.g., a pipe wall) and a flowing liquid or between a moving solid body (e.g., a rotating electrode) and a liquid constitutes a flow boundary layer or a hydrodynamic boundary layer within which the flow velocity is different from that of the bulk. This effect is known as viscous drag and is caused by the retardation effect of the solid surface (e.g., pipe wall) on the fluid. The thickness, δ_h, of the hydrodynamic boundary layer is sometimes defined as that distance from the wall at which the flow rate reaches 95% of the bulk value.[20]

If mass transport is occurring at the surface, there will also be a concentration boundary layer. In turbulent flow, convective transport (i.e., transport by the fluid velocity), due to the presence of turbulent eddies, will tend to eliminate concentration variations in the bulk of the fluid. However, turbulent fluctuations are damped near the wall, and in these regions, diffusion again becomes the dominant mode of transport. This region near the wall, in which the concentration of the reacting species differs from that of the bulk, is known as the concentration boundary layer. Its thickness, δ_d, is strongly dependent on velocity. As velocity increases, concentration variations are found to be confined to thinner and thinner boundary layers.

The relationship between the thickness of these two boundary layers is governed by the dimensionless Schmidt number (Sc), which compares the rate of transport by convection to that of diffusion. Convective diffusion analysis has empirically shown that:[21-23]

$$\frac{\delta_h}{\delta_d} \simeq (Sc)^{\frac{1}{3}} \simeq \left(\frac{\nu}{D}\right)^{\frac{1}{3}} \tag{1.5}$$

The Schmidt number ($Sc = \nu/D$) is not a function of flow velocity or of the characteristic dimensions of the flow geometry. It is determined exclusively by the physical properties that characterize momentum transfer and the transport of matter by a purely molecular mechanism.

When Sc is equal to unity (i.e., when $\nu = D$), there is a similarity between momentum and mass transfer. In a region of the liquid where the molecular mass transfer is controlling, molecular momentum transfer will also be expected to predominate.

Sc can be defined for both gases and liquids. For gases, the diffusion coefficient and the kinematic viscosity are numerically of the same order of magnitude, and thus $Sc = 1$. This is not the case for liquids. The kinematic viscosity of common liquids, such as water, is approximately 10^{-2} cm^2/s. The diffusion coefficients of molecules and ions in aqueous solutions are of the order of 10^{-5} cm^2/s and for macromolecules, $D \simeq 10^{-6}$ cm^2/s. Therefore, for water and comparable liquids, $Sc \simeq 10^3$. As the kinematic viscosity of the liquid increases, the diffusion coefficient decreases according to a relation of the type.

$$D = \frac{CONST}{\nu} \tag{1.6}$$

Such relations are only approximate.[24] Thus, Sc increases as the square of the kinematic viscosity. In viscous fluids, Sc may reach values of 10^6 and higher. A high Sc value means that, even at very low velocities, the transport of matter in a fluid by convection predominates over molecular diffusion. Molecular diffusion of matter is so slow that convective transfer of dissolved matter generally plays a more important role in liquids than in gases. The next sections will show the usefulness of Sc in mass transfer studies.

1.3 THE CONVECTIVE DIFFUSION EQUATION

The transport of ionic species in solution is by molecular diffusion (according to Fick's Law), by convection (i.e., entrainment with the fluid velocity), and by migration in an electric field. If the solution is an electrolyte of high conductivity such as sulfuric acid, or in the more general situation, when the solution contains a highly conductive supporting electrolyte, then electric fields are not involved in transport of the reacting species. In those situations, migration effects can be neglected, and the steady state transport of ions can be treated in the same way as an uncharged species, which is governed by the equation of convective diffusion:

$$\bar{v}\nabla c = D\nabla^2 c \tag{1.7}$$

This equation is applicable to laminar flow only, for in turbulent flow, the diffusivity (D) is no longer constant but, due to the effect of turbulent eddies, depends on velocity and position.

This equation may be solved in conjunction with the velocity distribution, $\bar{v}$, from a fluid mechanics treatment of the fluid flow by way of the Navier-Stokes equation.[25] This leads to a set of partial differential equations that are rarely amenable to analytical treatment.

The solution of such equations is in the form of a concentration profile across the liquid/solid interface, extending into the bulk of the fluid.[26] However, this approach is unwieldy because of the mathematical difficulties associated with it. For this reason, simplifying empirical approaches has been suggested as an alternative.

1.4 THE EMPIRICAL APPROACH TO CONVECTIVE DIFFUSION

The term "mass transfer" is used to describe the transport of ionic species by the combined effect of diffusion and convection, excluding the contribution of migration to the process. The mass transfer process is commonly referred to as "forced convection" when the system is mechanically agitated by stirring or pumping, and as "free convection" when mixing occurs as a result of density differences, accompanying concentration differences, from point to point within the fluid.

As a result of the contribution of diffusion to the transport process, it may be generally stated that movement of species towards or away from the electrode results in a decreasing concentration of the species in the direction of the transfer. Figure 1.1 shows examples of the two cases of cathodic and anodic control by diffusion of the reacting species.[15]

The exact shape of the concentration profile will depend upon the mass transfer mechanism predominating at any place. If one assumes that the concentration in the main body of the solution remains constant, it is then possible to express the overall mass transfer process by equations of simple empirical form that relate the rate of reaction to the concentration driving force, i.e., the difference between the average concentration of the transferred species in the main body of the solution and the concentration at the electrode-solution interface.

The mass transfer factor (K) is thus defined as:

$$K = \frac{\text{Rate of Reaction}}{\text{Concentration Driving Force}} \tag{1.8}$$

which can be rewritten in the case of an anodically controlled reaction as:

$$N = K(c_s - c_b) \tag{1.9}$$

The mass transfer rate (N) can then be related to the corrosion rate (CR) by a simple conversion between moles/cm^2/s and mm/y. The conversion factor for steels (A) is found to be $2.24 \cdot 10^9$,[9,12] so that

$$CR = AK(c_s - c_b) \tag{1.10}$$

In this way, Equation (1.10) can be used to predict corrosion rates when the reaction is diffusion controlled, and K, c_s, and c_b are known.

1.5 CORROSION RATE CONTROL

The mass transfer approach to the reaction rate described in the previous section is only applicable to reactions that are under anodic diffusion control or cathodic diffusion control. However, corrosion reactions encountered in practice can seriously depart from a simple and complete mass transfer control.

In general, the reaction at a corroding metal electrode is a series of elementary steps. These are:

- diffusion of the oxidant to the metal surface,
- reaction at the metal surface, and
- diffusion of the corrosion product away from the electrode.

The slowest of these steps will assume control of the reaction and will be one of the following:

- cathodic diffusion control,
- chemical or activation control at the anode or cathode, or
- anodic diffusion control.

Thus, a reaction under anodic diffusion control implies that it is occurring with negligible chemical polarization or that the surface reaction is infinitely fast compared to the anodic transport step. It also implies that the diffusion of oxidant to the surface is either very fast or the oxidant is in infinite supply.

Cases of mixed control have also been reported. An example of such a situation, using the RDE, is reported by Zembura (cf. also Section 2.4.1 and Figure 2.2), who studied the dissolution of electrolytic copper in air-saturated 0.1 N H_2SO_4.[27] At 25°C (77°F), chemical control was observed. At 50°C (122°F), the kinetics were intermediate; and at 75°C (167°F), the slow stage of the reaction was the diffusion of oxygen to the copper surface. The process of cementation at rotating zinc rods, which is shown to be diffusion controlled at low-to-moderate rotation speeds and activation controlled at higher velocities, is another example of mixed control.[28-30]

Studies by Eisenberg et al. of ferro-ferricyanide electrode reactions at rotating nickel electrodes have proved that the reactions may be considered to remain predominantly under mass transfer control up to a velocity of 157.3 cm/s.[31] However, chemical polarization is essentially present whenever finite currents are passed and become significant at high stirring rates when the surface reaction rate could become comparable to or slower than the transport rate. The reaction rate should then become independent of speed.

Also, an IR_f drop across the metal will occur if the corrosion product is deposited at the cathodic or anodic sites instead of diffused away and restricts or completely blocks contact between the metal and the solution. (R_f is the resistance produced by the film formed on the surface of the sites.) In the more general sense, there will also be a potential drop (IR_{SOL}) associated with the electrical resistance of the solution (R_{SOL}). In cases of solution of high conductivity, this can usually be neglected but may be significant in galvanic corrosion processes where the anodes and cathodes are separated by large distances.

The IR drop will contribute to the total overpotential, and there are specific situations where partial or total resistance control is experienced. As a consequence of the film constituting a barrier to the approach of reactants to one another, generally, the corrosion current will be reduced, and the corrosion potential may become more positive in cases of metals covered with protective porous films.[32] This applies particularly to the deposition of $CaCO_3$ and $Mg(OH)_2$ at the cathodic sites during corrosion in hard waters.[33-34] Since the anodic and cathodic sites are usually close together, the calcareous scale will also block the anodic sites and thus decrease the corrosion rate.

Another case of film control is found in the work of Mahato et al. on the corrosion of steel pipe in flowing aerated natural waters.[17] Initially, the corrosion is completely controlled by diffusion of oxygen, the cathodic reactant, to the surface. As corrosion proceeds, a buildup of corrosion products into a porous surface layer occurs. This in turn constitutes a second diffusional barrier that will eventually assume full control of the reaction rate and lead to a considerable reduction in the corrosion rate values and a reduction of sensitivity to flow rate variations.

All these types of polarization (activation, diffusion, and resistance) will be present to a greater or lesser extent in most corrosion reactions. However, should one be more significant than the others, it will control the rate of reaction. Only in those cases where diffusion is solely through the liquid boundary layer will the reaction rate be proportional to the concentration driving force and thus be described by Equation (1.9). In fact, the results of Mahato et al.[17] and Ross et al.[35] suggest that the situation could be greatly complicated by the appearance of surface films, the properties of which are sometimes difficult to determine.

1.6 THE LIMITING CURRENT DENSITY TECHNIQUE

1.6.1 The Experimental Observation of a Limiting Current

The cathodic polarization curve of a reaction controlled by diffusion of species towards the electrode of interest will show, upon slowly decreasing the applied potential, an initially sharp increase in current. This will then reach a saturation level as evidenced by a current plateau.

Figure 1.2 shows the experimental limiting current curve for copper deposition from an acidified cupric solution.[36] Only upon a relatively great further decrease of the applied potential will the current increase appreciably beyond this value and will then be due to additional electrode reactions taking place. For example, the appearance of gas bubbles at the cathode indicates that the electrode reaction is no longer exclusively the deposition of copper; hydrogen ions are also being discharged, resulting in the evolution of hydrogen gas.

This simple polarization experiment illustrates the basic features of the LCDT. A particular electrode reaction proceeds at the highest possible rate indicated by a current plateau. From the limiting current thus recorded, the mass transfer rate and the mass transfer factor at the electrode in question may be determined. This rate is determined by the composition and transport properties of the electrolytic solution and by the hydrodynamic conditions at the electrode surface.

Similarly, in the case of an anodically controlled reaction, a current plateau is observed upon increase of the applied potential.

1.6.2 The Derivation of the Limiting Current Density Expression

The molar flux of the reacting species is usually expressed using Fick's first diffusion law:

$$j = D\left|\frac{dc}{dy}\right| \tag{1.11}$$

The concentration gradient $\left|\frac{dc}{dy}\right|$ can then be rewritten (cf. Figure 1.1) as:

$$\left|\frac{dc}{dy}\right| = \frac{|c_b - c_s|}{\delta_d} \tag{1.12}$$

Furthermore, in concentrated solutions where the reacting species does not contribute noticeably to current conduction:

$$i = zFj \tag{1.13}$$

Combining Equations (1.11), (1.12), an (1.13), it is found that:

$$i = zFD\frac{|c_b - c_s|}{\delta_d} \tag{1.14}$$

The anodic diffusion-limited current density, i_L, can then be written as:

$$i_L = \frac{zFD(c_s - c_b)}{\delta_d} \tag{1.15}$$

Since the ratio $\frac{D}{\delta_d}$ is the mass transfer factor, K, this can be reformulated as:

$$i_L = zFK(c_s - c_b) \tag{1.16}$$

This is identical to the relation between mass transfer rate and concentration driving force previously found:

$$\frac{i_L}{zF} = N = K(c_s - c_b) \tag{1.17}$$

This relationship shows that a derivation of the limiting current density is possible if a value for K is known. (This will be the object of the next section.) It equally shows that the limiting current is the predicted corrosion current, I_{corr}, since Equation (1.9) related the mass transfer rate, N, to the corrosion rate within a proportionality constant, A, and that the limiting current measured from the anodic polarization curve is the actual corrosion current (cf. Section 3.3).

The surface concentration of the reacting species, c_s, is usually obtained from thermodynamic measurements and the bulk concentration, c_b, can be measured by titration.[3] The relationship is also the basis of the LCDT in mass and heat transfer studies whereby the mass transfer factor is arrived at from the measured limiting current density.[37-39]

In the introductory sections of Chapter Two, it will be shown how extensive use of the LCDT has led to the development of mass transfer correlations for a large number of systems of interest to engineers in mass transfer, heat transfer, and fluid mechanics. As a result of this wealth of information,[38] the technique has now found wide application in hydrodynamic corrosion research and related fields (e.g., electroplating) for the prediction of corrosion rates in diffusion-controlled systems.[9,12]

1.7 DIMENSIONLESS ANALYSIS IN MASS TRANSFER STUDIES

In the previous section, an expression (Equation [1.16]) for the limiting current density, i_L, has been derived. The calculation can only be implemented, allowing corrosion rate prediction (Equation [1.10]), if a way to calculate K can also be found. However, it is seldom possible to find the value of this factor by analytical calculations. In the majority of cases, it is necessary to make use of experimental data. Methods of generalization of these data thus become very important.

Ways to generalize experimental data are supplied by the similarity theory.[40,41] The theory postulates that any real physical phenomenon can be represented in the form of a relation between dimensionless numbers. Once this relation has been established, for certain geometric and physical conditions (by means of an evaluation of experimental data), it can further serve for the calculation of any process taking place under the same geometric and physical conditions but with different dimensions, velocities, and physical properties of the substance.

This has been recognized and developed for gases as long ago as 1874 by Reynolds.[42] However, it was Chilton and Colburn who extended the Reynolds analogy to liquids.[22] By an analysis of experimental data on heat transfer in liquids, they arrived empirically at what is today known as the Chilton-Colburn analogy between heat, mass, and momentum transfer. Expressed in its dimensionless form, it can be written as:

$$\frac{Sh}{Re\ Sc^{\frac{1}{3}}} = \frac{Nu}{Re\ Pr^{\frac{1}{3}}} = \frac{f}{2} \tag{1.18}$$

where the Nusset number (Nu) and the Prandtl number (Pr) are the heat transfer equivalent of the Sherwood and Schmidt numbers in mass transfer, and f is the friction factor. The Chilton-Colburn analogy reduces to the Reynolds analogy for gases whose Sc and Pr numbers equal unity. The analogy between the physical phenomena of mass, momentum, and heat transfer, though discovered empirically, has its origin in the similarity between their respective mathematical formulations; namely, Fick's Law for Diffusion, Newton's Law of Viscosity, and Fourier's Law for Heat Conduction. The mathematically based analogies are useful for two reasons. First, they can save mathematical work; if the solution to a heat conduction problem is known, the solution to the corresponding diffusion problem is also known. Second, mathematical analogies often suggest dimensionless groups that are in turn helpful in correlating the results of physical experiments.

The problem at hand is still the determination of the mass transfer factor, K. The similarity theory shows that it is possible to construct two different dimensionless numbers to describe mass transfer processes,[40] the first is the Sherwood number (Sh):

$$Sh = \frac{Kd}{D} \tag{1.19}$$

which is seen to describe the ratio of the rate of mass transfer to that of diffusion. The second number that can also be defined is the Stanton number (St):

$$St = \frac{K}{v} \tag{1.20}$$

St is a velocity-dependent mass-transfer dimensionless group. However, Sh is the most commonly used number in forced convection studies.

The mass transfer factor is a dependent variable, and the similarity theory, which is again helpful, has shown that in processes of forced convection, the Sh, for given geometric and physical conditions, is a definite function of the Re and Sc numbers:[40]

$$Sh = f(Re, Sc) \tag{1.21}$$

All that is needed at this stage to find the mass transfer factor, K, $\left(K = \frac{ShD}{d}\right)$ is to elucidate the form of the function in Equation (1.21). The theoretical determination of the form of Equation (1.21) that holds for the system of interest is based on the solution to the equation of convective diffusion. In view of the mathematical difficulties associated with it, (cf. Section 1.3), this approach is rarely tractable. When it is possible, however, the solution leads to an expression for δ_d, the diffusion boundary layer thickness, which for forced flow usually takes the form:[43]

$$\delta_d = \mathrm{CONST} \cdot d \cdot Re^{\frac{-m}{n}} Sc^{-\frac{1}{3}} \tag{1.22}$$

δ_d is introduced in Equation (1.15) for i_L, which then takes the form:

$$i_L = \mathrm{CONST} \cdot zF(c_s - c_b)\frac{D}{d} Re^{\frac{m}{n}} Sc^{\frac{1}{3}} \tag{1.23}$$

Equation (1.16) can also be written as:

$$\frac{i_L}{zF(c_s - c_b)} = K \tag{1.24}$$

Combining Equations (1.23) and (1.24), it is found that:

$$\frac{Kd}{D} = \mathrm{CONST} \cdot Re^{\frac{m}{n}} Sc^{\frac{1}{3}} \tag{1.25}$$

Comparing with Equation (1.19), the final result is arrived at:

$$Sh = \mathrm{CONST} \cdot Re^{\frac{m}{n}} Sc^{\frac{1}{3}} \tag{1.26}$$

The constants (CONST and m/n) in this equation will depend on the system of interest. The mass transfer factor, K, calculated from equations of this type can then be used to predict the corrosion rate (cf. Sections 1.4 and 1.6) by way of Equation (1.10).

For the rotating disc, the theoretical derivation of a mass transfer correlation of the form of Equation (1.26) preceded experimentation.[44,45] A semitheoretical treatment for turbulent flow at an RCE by Gabe and Robinson also leads to an equation of the same form (cf. Appendix A).[13] However, this expression is often obtained as an empirical correlation of experimental data obtained within certain limits. It is also obtained sometimes with the help of the Chilton-Colburn analogy for heat, mass, and momentum transfer.[46]

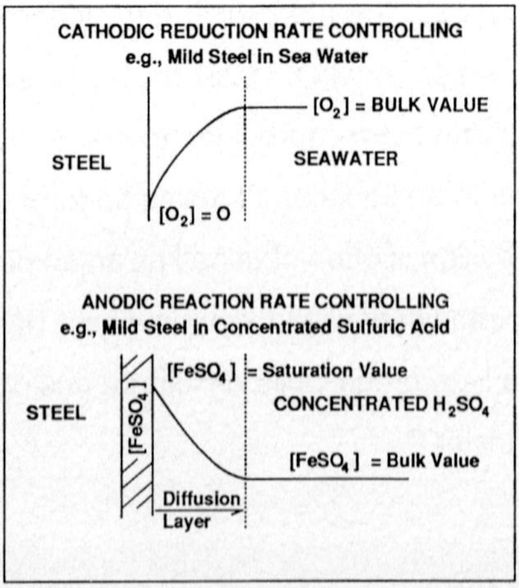

Figure 1.1 Examples of anodically and cathodically controlled corrosion reaction.[15] Reprinted with permission from *Corrosion Science*, vol. 23, B. Poulson. Copyright 1983.

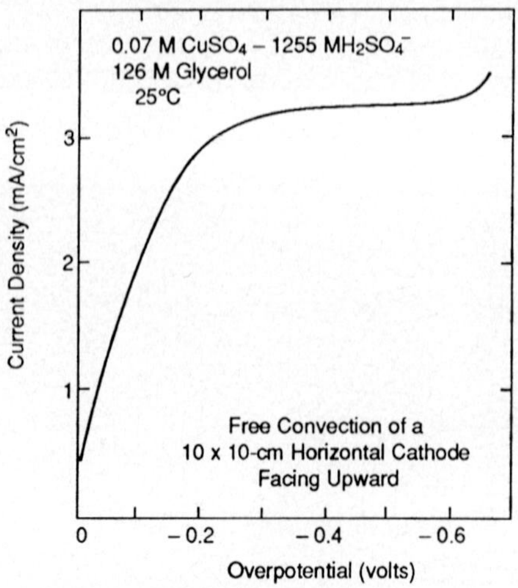

Figure 1.2 Experimental limiting current curve for copper deposition from an acidified cupric solution.[36] Reprinted with permission from R.G. Hickman.

CHAPTER
TWO

CRITICAL REVIEW OF HYDRODYNAMIC CORROSION RESEARCH USING MASS TRANSFER MEASUREMENTS

2.1 APPLICATION OF THE LIMITING CURRENT DENSITY TECHNIQUE (LCDT) IN CHEMICAL ENGINEERING RESEARCH

2.1.1 Introduction

The theory and practice of the limiting current density technique (LCDT) have been reviewed in Section 1.6. In recent years, wide ranging use has been made of this technique in fundamental chemical engineering research for the measurement of limiting electrolysis currents; in diffusion-controlled processes, it provides a versatile and accurate technique for the determination of mass and heat transfer rates.

The technique is also now widely used to characterize corrosion in flowing environments. This has only been made possible by the successful development of systems that can be operated under well-characterized, representative hydrodynamic conditions. J.R. Selman et al. have provided a list of mass transfer correlations obtained by means of the LCDT in over 200 investigations.[38] Parts of these studies were aimed specifically at developing a more rational basis for the design of electrochemical reactors. However, the greater part of the experimental work was undertaken by various investigators to exploit the unique advantage of the technique for obtaining local and average mass transfer coefficients over a wide range of hydrodynamic conditions. The primary interest was to gain insight into the nature of convective mass transport and also to provide practical correlations for the design and operation of process equipment. This highlights the fact that the approach is of interest, not only to engineers directly concerned with electrochemical processes such as corrosion and plating, but also to chemical and mechanical engineers involved in research in fluid mechanics, and heat and mass transfer.

2.1.2 The Electrochemical Reactions in Use

While LCDT is a very convenient means of continuously measuring mass transport, there are stringent requirements placed on the choice of electrode reaction.[19] The following criteria must be satisfied:

- The reaction is diffusion controlled.
- The reaction is stoichiometric, and the number of electrons (z) achieving discharge must be known and constant.
- The current efficiency is 100%.
- A limiting current density can be clearly measured from a potential-current density polarization graph.

Two electrochemical reactions predominate in experimental use; namely, cathodic deposition of copper and cathodic reduction of ferricyanide ions to ferrocyanide ions.

2.1.2.1 The Cathodic Deposition of Copper by the Reduction of Copper Ions

The cathodic reaction for reduction of cupric ions to metallic copper is given by:

$$Cu^{++} + 2e \rightarrow Cu$$

The electrolyte is usually aqueous $CuSO_4$, the supporting electrolyte is often 1.5 molar H_2SO_4, the test cathode surface is usually copper.[37] If the auxiliary electrode is also copper, so that the reverse reaction takes place on it, then the Cu^{++} concentration remains unchanged during a series of experiments. The main disadvantage of deposition processes is that the electrode surface on which deposition occurs may, in some circumstances, become roughened.[37] This could introduce complicating effects over and above the actual mass transfer studied.

2.1.2.2 *The Ferri-Ferrocyanide System*

Some oxidation-reduction reactions have also been suggested by several workers.[47-49] The advantage of using a redox reaction is that the electrode surface remains smooth throughout the experiment. One such reaction that is commonly used is the cathodic reduction of ferricyanide ions to ferrocyanide ions at an inert electrode according to the equation:

$$Fe(CN)_6^{3-} + e^- \rightarrow Fe(CN)_6^{4-}$$

The reverse reaction, oxidation of the ferrocyanide ion to ferricyanide ion, may occur at the anode so that again the bulk concentration of the electroactive species remains constant. The supporting electrolyte most commonly used is sodium or potassium hydroxide, and most workers have used an equimolar mixture of potassium ferro- and ferricyanide.[37] The most suitable electrode materials for this work are nickel,[46] platinum, and carbon. Two practical difficulties with this system are that exclusion of light and dissolved oxygen is necessary to maintain electrolyte stability.[37,46]

A longer plateau, giving better-defined limiting current, and the unchanging nature of the electrode surface geometry are the main advantages of this redox reaction over copper deposition.

Other diffusion controlled reactions that have been used are the oxidation of the ferrocyanide ion to ferricyanide ion, the reduction of quinone, and the reduction of dissolved oxygen.[50,51]

2.1.3 Systems for Measurement of Hydrodynamic Effects on Corrosion

A large number of flow systems have been more than adequately reviewed in the open literature. In *Electrochemical Measurements in Flowing Solution*, B. Poulson has systematically examined the apparatus used to study the effect of flow on corrosion.[15] For their part, J.R. Selman et al. list over 200 mass transfer correlations obtained for different geometries by means of the LCDT.[38] The reported studies are both theoretical and empirical in nature and cover both laminar and turbulent regimes.[38]

Several examples of typical geometries used for corrosion assessment are reviewed. These include a submerged jet impinging on a flat surface,[52] orifices and nozzles to reproduce the severe hydrodynamic conditions that lead to erosion-corrosion,[53,54] plates in duct walls, circular concentric annuli, and multiple plates in pipes.[15]

The rotating disc electrode (RDE) is the most widely used system in electrochemical studies and analytical chemistry and is unique in that theory preceded experimental studies. Its main advantage is that laminar conditions are maintained over a wide range of velocities. The exact solution to the theoretical equations has been derived for mass transfer under laminar conditions by Levich.[44] Modifications have been proposed by Sparrow and Gregg,[55] Riddiford,[56] and Newman.[45] However, the original Levich treatment is adequate for use in corrosion studies.[44,15]

The rotating cylinder electrode (RCE) because of its convenience, and pipe flow because of its wide applicability, have also been thoroughly investigated in numerous studies. This is particularly true in corrosion research as is indicated in Section 2.4. Some of the published research on both geometries is treated in detail in the following sections.

2.2 THE CONCENTRIC ROTATING CYLINDER

2.2.1 The Early Years

Earlier workers on the RCE geometry were only concerned with the achievement of a particular rotation speed. They used the rotating electrode primarily as a means of stirring their electrolyte and paid little attention, neither to geometrical considerations nor to the nature of the reaction, to allow them to have well-defined hydrodynamic and reproducible effects.

The first study of the effect of the speed of rotation upon the rate of mass transfer was conducted by Brunner who found that:[57]

$$\delta_d = \frac{0.05}{R^{\frac{2}{3}}}$$

where δ_d is the diffusion boundary layer thickness (cm) and R is the rotation speed (rpm). Rotation speeds of up to 150 rpm were used for this work. However, he considered neither the effect of rotor diameter nor the dependence on the physical properties of the electrolyte. The result, therefore, could only be given qualitative importance.

Roald and Beck investigated the dissolution of rotating magnesium rods in hydrochloric acid solutions.[58] They found that the reaction rate increased with the speed of rotation to a power of 0.71 for low-acid concentrations. At acid concentrations greater than 1.4 M, the reaction rates became independent of rotation speed. The evolved hydrogen bubbles apparently contributed more strongly than the rotation to the forced convection. For this reason, the work of Roald and Beck represents a special case of mass transfer.

2.2.2 The Work of Eisenberg, Tobias, and Wilke[46]

Eisenberg et al. were the first to produce a comprehensive study of mass transfer at a rotating electrode. Their primary aim was: "To establish correlations between the physical, geometrical and hydrodynamic conditions and rates at which a solute (ion) is transferred to or from a rotating electrode."

The apparatus used by Eisenberg et al.[46] fulfilled the requirements of hydrodynamics and potential theory (electrostatics) as outlined by Karlberg.[20] These are as follows:

> For the determination of anodic and cathodic polarization curves of corroding metals, the applied current should be distributed as uniformly as possible across the surface studied. An electric current tends to concentrate at corners and edges of an electrode. From the viewpoint of potential theory (electrostatics), the ideal arrangement would be two concentric spheres. Such an arrangement is, however, difficult to achieve practically. The second best geometry, according to potential theory, is two coaxial cylinders. This geometry is found to be a practical hydrodynamic measurement system. An arrangement consisting of a rotating cylinder, surrounded by a coaxial stationary cylinder, offers great advantages from the viewpoints of both hydrodynamics and potential theory. The peripheral velocity is constant, which means that both the diffusion layer thickness, δ_d, and the shear stress between the electrode and the electrolyte are constant across the circumferential surface of the cylinder. The current distribution can be made practically uniform, partly by insulating the basal surfaces of the rotating cylinder, and partly by making the height of the cylindrical electrodes almost equal to the height of the electrolysis vessel, whose bottom and lid then act as efficient current shields.

The nature of the flow between concentric cylinders, the inner of which is rotating, has also been investigated by Eisenberg et al.[46] In special experiments involving injection of dyes into the fluid during the rotation, fully developed turbulence was observed at the lowest speeds used in their studies. It has been found that the critical Reynolds number (Re_{CRIT}) for transition between laminar and turbulent flow is very small, $Re_{CRIT} \approx 200$, corresponding to rotational velocities of less than 10 rpm.[19,46] The RCE can then be said to operate in the turbulent range at all but the very low velocities. This is one of the main advantages of the RCE that makes its use in corrosion studies highly desirable because it can be used to simulate pipe flow in plant conditions that are usually turbulent (cf. Section 5.1.6.1). The turbulent nature of the flow in the RCE is also of great interest in electrodeposition where it allows high rates of deposition to be achieved (cf. Section 2.2.3). At the low rotational velocities necessary to secure laminar flow, the contribution of natural convection to the rate of mass transfer would obscure that of forced convection produced by rotation. Thus, in order to make negligible the effects of a superimposed natural convection, all studies using the RCE should be conducted under conditions of well-developed turbulence. This suggests that the RCE may be of little use in laminar flow conditions.

Eisenberg et al. used an apparatus that incorporated an outer cylinder auxiliary electrode that totally surrounded the rotating electrode and also acted as the electrolyte vessel. Also, once the RCE was mounted securely, the cell was completely filled with the electrolyte in order to avoid air bubble formation and eliminate rotational vortices.

In addition, their research pointed out the usefulness of the rotating cylinder in maintaining a diffusion layer of uniform thickness that in turn allowed a uniform current distribution over the entire surface.

The study paid particular attention to the choice of electrode reactions investigated. The requirements imposed on an electrode reaction used to study mass transfer rates by the LCDT are detailed in Section 2.1.2. The considerations led to the selection of the cathodic reduction of ferricyanide to ferrocyanide at an inert (nickel) electrode and the reverse reaction, the oxidation of ferrocyanide:

$$Fe(CN)_6^{3-} + e^- \underset{\text{light}}{\overset{\text{dark}}{\rightleftharpoons}} Fe(CN)_6^{4-}$$

These reactions are known to occur with negligible chemical polarization and, thus, can be used to obtain mass transfer rates by measurement of the limiting current observed on polarization of the electrode.[31] Another advantage in using a redox reaction is that the electrode remains smooth throughout the experiment.

They also included in their investigation dissolution studies of solids cast into a rotating cylinder; namely, benzoic acid dissolution in water, 3.65 M glycerol-water, and also 4.44 M glycerol-water.

The inner and outer cylinder diameters, the viscosity and diffusion coefficient, and the rotating speed of the inner cylinder were varied so as to give a range of property values to be studied. This is summarized in Table 2.1.[46]

They were able to correlate the results obtained using the empirically derived Chilton-Colburn analogy between heat, mass, and momentum transfer.[22] Expressed in its dimensionless form, the Chilton-Colburn mass transfer factor, j_D, can be written as:

$$j_D = \frac{Sh}{Sc^{\frac{1}{3}} Re} = St \cdot Sc^{\frac{2}{3}} = f(Re)$$

The analysis of their experimental results led to the following findings:

- The characteristic dimension in the Reynolds number (Re) was found to be the rotor diameter, d, rather than the gap, h, between the cylinders, as was previously reported by Taylor.[59]

- The mass transfer data were correlated by the following relation:

$$St \cdot Sc^{0.644} = 0.0791\, Re^{-0.3}$$

which, reverting to the more commonly used Sh, reads as:

$$Sh = 0.0791\, Re^{0.7} Sc^{0.356}$$

This relation successfully correlated, within experimental error, all the results obtained for the five electrochemical systems and the ranges of Re and Schmidt (Sc) numbers used. However, the cylinders were noticeably roughened during the dissolution experiments giving slightly higher rates than obtained from the smooth cylinder-redox reaction.

This point was studied further by Makrides et al. who conducted a least square analysis of the results published by Eisenberg.[60] They found that the velocity dependence exponent was higher for oxidation (0.77) than reduction (0.72). It was also seen to vary with the Re range. The value was 0.68 for the interval Re = 1,000 to 10,000, while it was 0.77 for Re = 30,000 to 100,000.

In conclusion, the Eisenberg et al. study provided a useful correlation of mass transfer at a rotating cylinder. However, no mechanistic explanation was proposed by the authors.

2.2.3 The Experimental Studies of Gabe and Robinson

Gabe and Robinson studied high-speed electrodeposition of copper from copper sulfate solutions onto a rotating copper cathode.[61,62] The outer cylinder was used as the electrolyte container and anode. Their work is highly relevant to both metal finishing and extractive metallurgy fields where the use of high-current densities may lead to a higher output and a reduction in operating costs on capital plant. Their reported mass transfer correlation was:

$$Sh = 0.079\,Re^{0.69}\,Sc^{0.41}$$

The value of the Re exponent agrees remarkably well with other investigations, although the Sc exponent is inexplicably high. Such discrepancies between values reported by different authors may be due to differences in cell dimensions and the fact that only a limited range for Sc could be examined owing to the limitations set by solution composition considerations.[62] Gabe and Robinson were also responsible for a semitheoretical derivation of a mass transfer relation describing turbulent flow at an RCE (cf. Appendix A).

2.2.4 Discussion of Other Work Using the Rotating Cylinder Electrode (RCE)

Following the Eisenberg studies, Arvia and Carroza, investigating the deposition of copper from a copper sulfate electrolyte, obtained an empirical relation for mass transfer to the outer stationary cylinder.[63] The apparatus used was carefully designed and fulfilled the requirements of hydrodynamics and potential theory.[20]

The gap between the inner rotating electrode and the outer stationary cathode was varied by using different rotor diameters. A range of physical properties of the solution was also studied. The correlation was reported to be:

$$Sh = 0.079\,Re^{0.7}Sc^{0.356}\left(\frac{d_1}{d_2}\right)^{0.7}$$

where d_1 and d_2 are the outer and inner cylinder diameters, respectively.

Newman has also suggested that the ratio of inner to outer cylinder diameter should be incorporated in the correlation,[64] thus modifying the Re term to:

$$Re^{0.7}\left(\frac{d_R}{d_L}\right)^{0.7}$$

where d_R is the diameter of the inner rotating cylinder, and d_L is the diameter of the cylinder with the limiting current. In the work of Eisenberg et al., the limiting electrode was the inner, rotating electrode with $d_R = d_L$. Eisenberg et al. reported a noticeable roughening of the surfaces in their dissolution experiments. At the same time, an increase in the velocity dependence exponent was observed.[60]

Makrides and Hackerman, and King and coworkers report that the mass transfer varies linearly with the angular velocity in certain cases of cylinder dissolution.[60,65,66] Makrides and Hackerman considered this result to be due to the cylinders becoming rough during dissolution.[60] They were the first to recognize and explain this important phenomenon. R. Kappesser, I. Cornet, and R. Grief noted that the use of a dissolving surface by Eisenberg et al. does introduce time variations in the surface roughness.[67,46] Due to these considerations, as well as the desire to extend the applicability of the results of Eisenberg et al. to a wide range of Re and roughnesses, Kappesser et al. utilized another mass transfer system to study this problem.

The apparatus consisted of a vertically mounted, nondissolving cylinder that was rotated in an oxygenated 4% NaCl-H_2O solution. The mass flux of the oxygen to the cathodic cylinder surface was obtained from the mass transfer limiting current for the reaction:

$$O_2 + 2H_2O + 4e^- \rightarrow 4OH^-$$

The test cylinders were made of Monel[(1)]. Four cylinders were used, the first was a highly polished cylinder with a surface roughness of less than $3 \cdot 10^{-5}$ cm rms height.

[(1)] Trade name.

The remaining cylinders had staggered diamond knurls machined on their surfaces. The knurls had peak-to-valley height (h) of $7.3 \cdot 10^{-2}$ cm (cylinder 2), $4.1 \cdot 10^{-2}$ cm (cylinder 3), and $1.52 \cdot 10^{-2}$ cm (cylinder 4) giving d/h ratios of 87, 156, and 416, respectively. The diameter of the cathode was d = 6.35 cm; the polished cylinder had a d/h ratio greater than $2 \cdot 10^5$.

They were able to correlate their results using the following mass transfer correlation:

$$Sh = \left(1.25 + 5.76\log_{10}\frac{d}{h}\right)^{-2} Re\,Sc^{0.356}$$

Fig. 2.1 shows their results plotted as $Sh/Sc^{0.356}$ vs Re for all four cylinders used.

2.3 THE PIPE FLOW

2.3.1 The Empirical Derivation by P. Harriot and R.M. Hamilton of a Mass Transfer Relation for Turbulent Pipe Flow[68]

The method used by these workers consisted of calculating rates of transfer from the loss in weight of soluble pipe sections through which liquid was circulated. The solid was benzoic acid, and the liquids were water, glycerine-water solutions, and methanol-water solutions.

The test cylinders were pressed from reagent grade benzoic acid in a polished metal mold as suggested by Meyerink and Friedlander.[69]

The test section was preceded by 178 cm of straight plastic pipe. This allowed established mass transfer conditions to be obtained before the specimens were reached.[70] It is also necessary to provide sufficient length of straight pipe ahead of the specimen so that the hydrodynamic boundary layer may be fully established. Ross and Jones summarized these requirements as being:[71]

- for laminar flow $L_{MIN} = 0.035\ Re \cdot d\ (Re<2{,}000)$
- for turbulent flow $L_{MIN} = 0.693\ Re^{0.25} \cdot d (Re>3{,}000)$

The viscosities and diffusion coefficients of the solutions used allowed the Re to vary between 10,000 and 100,000 and the Sc from 430 to 100,000.

Examination of the used cylinders showed no noticeable roughening, and consequently, the mass transfer characteristics observed are typical of a smooth turbulent pipe flow.

P. Harriot and R.M. Hamilton found that their data were correlated with an average deviation of 5.4% by the equation:

$$Sh = 0.0096\,Re^{0.913}Sc^{0.346}$$

The above mass transfer correlation gives mass transfer data 5 to 10% below the correlation of Friend and Metzner and 20 to 100% above other published correlations.[72-74]

2.3.2 Corrosion of Steel Pipe in Concentrated Sulfuric Acid

The method used by Ellison and Schmeal for determining mass transfer coefficients is essentially that of Harriot and Hamilton.[9,68] Their test sections consisted of pipe specimens mounted in a straight length of the pipe-line apparatus with due care given to the entry length requirements.[9]

The data were obtained in flowing 68% sulfuric acid at 27°C (81°F), 40°C (104°F), and 60°C (140°F). Twenty-four hour weight data were taken on the pipe specimens and then formulated into a Sh.

The results were compared with mass transfer correlations proposed by several investigators.[68,72-74] The data were found to fit best the correlation of Harriot and Hamilton for the condition tested.[68]

$$Sh = 0.0096\,Re^{0.913}\,Sc^{0.346}$$

The measured corrosion rates agree with the mass transfer correlation to within 5%.

2.4 HYDRODYNAMIC CORROSION RESEARCH REVIEW

2.4.1 Introduction

An extensive amount of laboratory corrosion research reported in the scientific and technical literature is done under quiescent conditions. In real-life, industrial plant operation, the corrosive fluids are often pumped around pipes, heat exchangers, and vessels. The simple fact that fluid flow often affects the corrosion rate and mechanism of an alloy in a given environment makes the use of corrosion results obtained in laboratory beakers, stirred or agitated under poorly defined conditions, questionable if not misleading or erroneous. This inaccuracy was recognized long ago, and test methods have been developed to quantitatively define the effect of fluid velocity. Among these methods, the laboratory flow loop and the RCE have been most widely used.

The laboratory flow loop with its well-defined hydrodynamics should be able to simulate pipe flow in the plant situation in a quantitative fashion; however, it suffers from some major drawbacks. Large amounts of fluid are required. This is particularly demanding and hazardous when such fluids are either highly corrosive or even toxic. It is very expensive and time consuming because it requires the installation of pumps, temperature control equipment at a number of locations, flow measurement devices, and extensive leak containment provisions. In view of these limitations, the RCE with its well-defined hydrodynamics, ease of assembly and disassembly, smaller volume of fluid used, and easier flow and temperature control, has been preferred by a number of workers.[9,13]

Dimensionless analysis using mass transfer concepts showed that the corrosion, when controlled by diffusion of one of the species between the bulk fluid and the surface, could be modeled completely by the rate of mass transfer of the rate limiting species and the Re and Sc numbers.[9,12,35] In general, the effect of flow can be used to determine if corrosion is under activation, diffusion, or mixed control.

An excellent example of flow as a diagnostic technique for the type of corrosion mechanism is shown in Figure 2.2, from the work of Zembura, in which the corrosion of copper discs rotating in air-saturated 0.1 N sulfuric acid at different temperatures is examined (cf. Section 1.5).[27]

In the following sections, some aspects of hydrodynamic corrosion research as reported in the open literature, particularly in a paper by B. Poulson, will be reviewed.

2.4.2 Uniform Corrosion

Mathematical modeling using mass transfer concepts has been successfully applied to a number of cases of uniform corrosion. Examples are the studies of Ross, Wood, and Mahmud of the corrosion of steel in sulfuric acid,[35] and that of Mahato, Cha, and Shemilt of the corrosion of steel pipes in aerated natural waters.[17] In these cases of uniform corrosion, the mechanism has often been one in which the corrosion rate either is determined by, or greatly influenced by, the convective diffusion rate of one of the species between the surface and the bulk liquid.

In a major study, Ellison and Schmeal used the RCE technique to investigate the corrosion of steel in 68% and 78% concentrated sulfuric acid at three temperatures; namely, 27°C (81°F), 40°C (104°F), and 60°C (140°F); and over the 0.4 to 1.6 ms^{-1} velocity range.[9] They found that the corrosion rate was controlled by the transfer rate of ferrous ion from a ferrous sulfate saturated layer-liquid interface into the bulk acid.

They were then able to correlate their result using the mass transfer correlation previously derived by Eisenberg:

$$Sh = 0.079 Re^{0.7} Sc^{0.356}$$

with a ± 25% error margin.[46]

Some other cases where corrosion is diffusion controlled include copper alloys in rapidly flowing water.[75] In this case, the solubility of the copper corrosion product is the likely limiting step. However, the analysis is complicated by the variety of corrosion products that can form on copper alloys. Similarly, Mahato et al. and Ross and co-workers observed complicating effects due to film formation (cf. Section 1.5).[17,35] Another example of mass transfer limited corrosion is the case of carbon steel in carbonic acid solutions with concentration of dissolved carbon dioxide between 0.1 and 1%.[76]

2.4.3 Passivation

It is postulated that there are two mechanisms of passivation:

(1) dissolution followed by precipitation, and
(2) solid state reaction (oxide or adsorption) in parallel with metal dissolution.

Possible criteria for deciding between the two have been examined by Armstrong, who suggested that the active-passive transition should be rotation (flow) dependent if mechanism (1) occurs, as long as the nuclei are formed at a distance from the electrode where they can be swept away.[77] No effect should occur if (2) were the operating mechanism. Turner and Brook used this technique to examine the behavior of copper in hydrochloric acid solutions and found no effect of flow on the formation of CuCl.[78] They were able to conclude that it formed by a solid state mechanism. Wilde has shown the reverse situation studying the behavior of iron in 1 M LiCl.[79] In the absence of flow, no passivation occurs because of acidity building up. This is removed by using a rotating cylinder or pipe flow, and passivation is then able to occur.[15]

2.4.4 Pitting

A number of models have been postulated for the breakdown of passivation and the initiation of pitting. One particular model suggests that passivity breakdown is associated with the localized build-up of acidity at the metal surface. Some workers have taken up the idea that this could be tested by examining the effects of fluid flow on the breakdown potential. This is summarized in Table 2.2, taken from Poulson.[15]

It would appear from the table that the flow conditions examined have little effect on the pitting potential. However, the ultrasonic vibrations have a noticeable effect.

The growth of pits has been shown to be minimized by the influence of solution flow. This has been pointed at by the general recommendation that turbulent flow conditions corresponding to velocities greater than 5ft/s are required to avoid pitting of stainless steel in seawater.[84]

Beck and Chan have examined the effect of flow on the growth of pits on iron and commercial type 304 (UNS S30400) stainless steel in sodium chloride solutions.[85] They found that a salt film is more important than low pH in maintaining pitting. They also found that above a critical velocity, the salt film is removed and repassivation occurs. They developed a theory suggesting that such processes depend on the size of the pit, and the higher the velocity, the smaller will be the critical pit size at which repassivation occurs.

More recently, Silverman published a study that proposed a systematic approach to solving problems of localized corrosion under flowing conditions.[86] He suggested that the dimensionless analysis successfully used for modeling uniform corrosion can be a fruitful approach. New dimensionless numbers arise from modeling the localized area. These are the ratios of the magnitudes of migration to diffusion, convective mass transport to diffusive mass transport (modified Peclet number, $Pe = Re \cdot Sc$), and surface reaction rate to mass transfer rate (modified Dahmkoler number). These dimensionless groups, along with Sh, can provide guidance for experimental design and corrosion prediction.

Table 2.1

Range of Property Values Studied by Eisenberg et Al.[46]

System	Rotor Diam. d(cm)	Gap-to-Rotor Ratio h/d	Peripheral Velocity, cm/s	Reynolds Number (Re)	Schmidt Number (Sc)
Benzoic Acid into H_2O	1.94 to 5.98	0.350 to 2.97	1.06 to 200.8	239 to 241,600	835 to 975
Benzoic Acid into 3.65 M glycerol-H_2O	2.02 to 4.30	0.610 to 1.99	8.09 to 362.5	786 to 41,700	4,926 to 5,053
Benzoic Acid into 4.44 M glycerol-H_2O	4.20 to 4.31	1.36	11.6 to 354.6	1,636 to 48,800	10,847 to 11,491
Oxidation of ferrocyanide	1.27 to 5.02	0.104 to 3,377	1.17 to 426	112 to 162,000	2,595 to 3,649
Reduction of ferricyanide	1.27 to 5.02	0.104 to 3,377	1.17 to 426	112 to 162,000	2,232 to 3,134

Table 2.2

Summary of Effects of Flow on Pitting Potentials[15]

Metal	Environment	Specimen Type	Effect	Reference
Aluminium	3.5% NaCl	Rotating cylinder	None	Mansfield[80]
Iron	1 M LiCl	Rotating cylinder	None	Wilde[79]
Aluminium	0.1 N NaCl	Rotating cylinder	Increase by 50 mV	Franz et al.[81]
304L (UNS S30403)	0.2 M NaCl + 0.1 Na_2SO_4	Rotating cylinder	Almost no effect	Sato et al.[82]
304	0.1 N NaCl	Flat specimen in solution vibrated with 200 kHz waves	200 mV more noble	Nakayama[83]

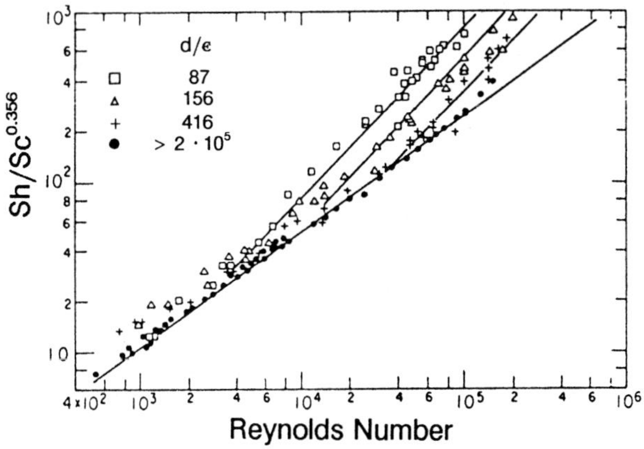

Figure 2.1 Mass transfer rate at rough/smooth cylinders plotted as $Sh/Sc^{0.356}$ vs Re.[67] Reprinted with permission of the publisher, The Electrochemical Society, Inc.

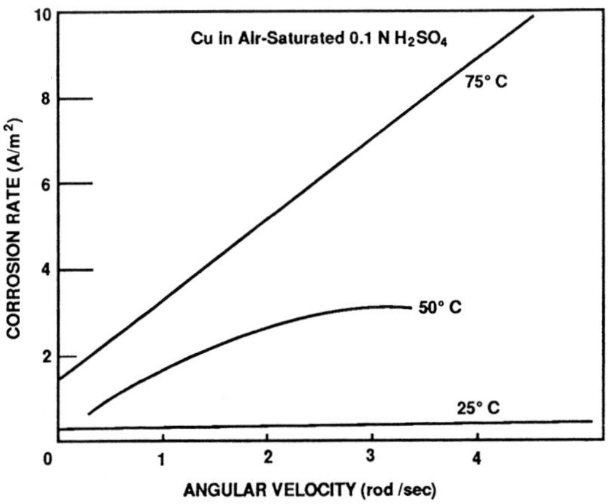

Figure 2.2 Effect of angular velocity and temperature on Cu discs rotating in air-saturated 0.1 N H_2SO_4.[27] Reprinted with permission from *Corrosion Science*, vol. 8, Z. Zembura. Copyright 1968.

CHAPTER
THREE

EXPERIMENTAL

3.1 DESCRIPTION OF THE APPARATUS

The rotating cylinder electrode assembly (RCE) is described schematically in Figure 3.1, while Figure 3.2 is a photograph of the entire experimental setup. The RCE assembly was mounted in a rigid structure specially designed to minimize mechanical vibration. This, in turn, is supported on a concrete floor by a frame, the legs of which have levelling screws and antivibration pads.

A stepper motor was used to drive the rotating shaft. Transmission of the rotation to the shaft was made by a belt and pulley system. The motor was operated from a regulated 220 V AC voltage supply and controlled by a servo-amplifier loop for optimum stability of the desired rotation rate to within ±5 rpm. The shaft could be driven in this way up to a maximum rotation of 10,000 rpm.

The 8-way slip-ring assembly was used to read the electrical signals from the electrode and was firmly mounted at the upper end of the rotating shaft. The slip-ring unit was secured rigidly to the RCE supporting structure by means of four mounting bolts. The unit rotor was then driven through a flexible coupling from the rotating shaft. The coupling's flexibility was sufficient to prevent loading of the slip-ring bearings by any eccentricity or end-float of the rotating shaft.

The rotating movement was transmitted to the lower part of the shaft by a set of four high-speed shielded ball bearings to achieve minimization of eccentric motions at the lower end of the electrode. The rotating electrode was screwed on to the lower end of the rotating shaft, which was tapered to minimize further eccentricity.

The slip-ring unit was fitted with silver graphite brushes. Continuous silver connections were also used throughout the unit, thus giving satisfactory electrical contact throughout the duration of this study. With the grade of material necessary for the application, some brush wear was inevitable, and periodically the inspection cover was removed and the silver graphite dust cleaned away. To prevent any dust being deposited on the brush block between the springs, a soft, long-haired brush was used.

The RCE was fabricated from AISI 316L (UNS S31603) stainless steel. Its diameter was 31 mm, and it had a length of 210 mm. The test specimens, 31 mm in diameter and 10 mm in length, giving an approximate area of 9.74 cm^2, were sleeve-slipped on a polytetrafluoroethylene (PTFE) insert to achieve electrical insulation between the shaft and the specimens. Furthermore, the shaft was isolated from its supporting structure by way of a polypropylene insert. A similar insert was used to achieve insulation between the slip-ring assembly housing and the RCE supporting structure. The test section can accommodate up to five specimens. Electrical isolation between the different specimens was achieved by way of PTFE spacers.

Electrical connections were made between the electrodes and the appropriate pins on the slip-ring unit using screened wires. The connection was made by eight BA set screws into tapped holes on the inner side of the electrodes. The common screen was then connected to the Autostat[(1)] case. The RCE supporting structure and the bare metal surfaces of the instruments used were then connected to a common ground to eliminate electrical noise pick-up from the stepper motor, the noise from the mains, and airborne noise. Signals to the recording instruments were carried from the brush block end of the slip-ring unit assembly.

With the electrode specimens and spacers mounted on the RCE, the stainless steel spindle of the electrode was tightened to secure the probe assembly into position. The tightness was sufficient to prevent crevice formation without gross deformation of the PTFE spacers. It is important that the PTFE spacers are flush with the electrode specimens in the ready-to-use electrode assembly.

A 316L stainless steel ring, with the same dimensions as the test pieces, was used as the auxiliary electrode. This was electrically connected to the measuring instruments in the manner already described

---

[(1)]Trade name.

for the working electrode. The decision to use this geometry rather than a concentric cylinder has received full justification (cf. Sections 4.2.2 and 5.1.4). Two separate and identical working electrodes were used throughout the present series of experiments.

The acid was contained in a glass cell, 160-mm internal diameter and 300 mm in height, capable of holding up to approximately 6 L of solution. The top surface of the cell was ground flat. A schematic diagram of the cell is given in Figure 3.3. The size of the cell allows experiments of reasonable duration to be conducted without the necessity of continuously renewing the acid. Acid replenishing may be necessary from time to time as ferrous ion contamination increases following corrosion. Successive efforts to seal the cell were abandoned when a graphite-filled, spring-loaded PTFE running seal, of a type specially designed for the sealing of chemical pumps, failed to seal the cell successfully. The problem was in the friction between the cell and the stainless steel rotating electrode, leading eventually to an opening up of the seal. Nevertheless, the extent of bubble formation and vortex shedding was only significant at the higher velocity, and the experiments were allowed to proceed. Acid was introduced into the cell through a specially designed inlet in the cell top. The cell top was securely mounted to the supporting frame, contained the housing for the seal, and had provisions for an overflow system. As a consequence of the use of the cell top, the acid inlet on the glass cell became disused. The inlet on the cell top was also used as an access port for a gas inlet pipe to provide a gas blanket of oxygen-free nitrogen over the acid surface. The salt bridge was filled with a potassium sulfate-doped polyacrylamide gel, prepared according to the method described in Appendix B. The gel, renewed regularly, gave satisfactory results throughout this study. Contact between the gel and the Hg/Hg_2SO_4 reference electrode used was made by way of a solution consisting of 6 g potassium sulfate in 50 mL distilled water contained in a Quickfit[(2)] holder. This was connected to the salt bridge by a flexible tube. No attempt was made to correct the measured potentials to include the small temperature variations of the reference potential.

The cell rested on a laboratory jack so that it could be raised or lowered as necessary during the course of the experiments. It was also secured to the base of the supporting structure of the RCE by four rods fastened on a supporting plate. It was precisely located into a circular PTFE-lined groove machined into the cell top.

Satisfactory temperature control was achieved using a recirculatory water system through the built-in water jacket designed to surround the cell. The system was completed by a model 5-MD pump[(3)] and a thermostatically controlled water bath to heat the water. Using this system, it was possible to keep the acid temperature constant to within $\pm 1^{\circ}C$ ($34^{\circ}F$).

The anodic polarization experiments and the corrosion potential measurements were made using a software-controlled potentiostat, namely, an Autostat controlled by a 32K microcomputer.

[(2)]Trade name.

[(3)]Little Grant Corporation, Oklahoma City, Oklahoma, United States of America.

3.2 MATERIALS USED

Prior to each experiment, the specimens were finished to 1,200 grade emery paper, degreased, and then weighed. After each experiment, the specimens were removed, washed in distilled water, and then abraded lightly with a rubber eraser to remove any remaining sulfate film and loose corrosion products. The specimens were then reweighed.

Analar[(4)] sulfuric acid (analytical grade) was used exclusively for the present work. The nominal concentration of the acid supplied was 98% w/w; its iron content was 0.0001%. (The w/w symbol designates concentration given by weight; however, for the purposes of this investigation, the w/w denomination will be omitted.) Ninety-eight percent sulfuric acid was mixed with distilled water to make 93% and 68% sulfuric acid. Dilution was conducted by volume.[87]

A series of four cast irons, namely, grey cast iron, ductile iron, Mondi A, and Mondi B, was tested in 98% sulfuric acid at 27°C (81°F) and 40°C (104°F) over the 0.81 to 2 ms^{-1} velocity range. The chemical analysis of the alloys is given in Table 3.1.

The mild steel used was AISI 1020 (UNS G10200). It was successively tested in 68%, 93%, and 98% sulfuric acid at 27°C (81°F) and 40°C (104°F) and over the 0.81 to 2 ms^{-1} velocity range.

[(4)]Trade name.

3.3 EXPERIMENTAL PROCEDURE

The study of corrosion of carbon steel in concentrated sulfuric acid shows the anodic polarization curve to have a vast region of potential where the current is virtually independent of potential. The potentiodynamic polarization measurements were made using an Autostat controlled by a 32K microcomputer with a resolution of 0.076 mV on potential and 0.06 μA on current. The standard procedure adopted when carrying out a polarization curve was as follows:

- Measure the corrosion potential of the working electrode.
- Set the potential of the Autostat to the corrosion potential.
- Scan anodic from the corrosion potential to the anodic limit (typically approximately -100mV)
- Scan cathodic to the corrosion potential.

The scan rate for the measurement was generally 300 mV/h. The current-potential data were then stored as a separate file for each polarization curve on a floppy disc for further analysis.

One of the earliest studies of the anodic polarization behavior of mild steel in concentrated sulfuric acid was conducted by Hines and Williamson.[88,89] They recognized two distinct portions to the polarization curve. The first one is the sulfation range where current is independent of potential. It was found in this study to extend from about +100 mV above the corrosion potential to a potential of approximately 0 to 100 mV with respect to the Hg/Hg_2SO_4 reference electrode (cf. Figure 3.4). Hines and Williamson found the film formed in this region to be duplex: the inside layer being a thin, compact film and the outside layer a thick, porous ferrous sulfate film. The second distinct region appearing in the current potential diagram (Figure 3.4) is the passive range, greater than about 0 to 100 mV with respect to the Hg/Hg_2SO_4 electrode, where the film is apparently covered by a ferric sulfate film over a passive oxide film.[89,90] The sharp peak at about 0 mV in Figure 3.4 results from the change in oxidation states of the iron sulfate film already existing on the metal surface. The findings of Hines and Williamson and of Ellison and Schmeal, with respect to the general shape of the polarization curve and to the critical potential values, have been confirmed in this study (cf. Figure 3.4).[9,88,89]

The independence of anodic current on potential implies that the anodic current, in the sulfation range, is also equal to the free-corrosion rate. That is, the anodic process limits the corrosion rate. The distinct current plateau observed in this system allowed an easy determination of the experimental value of the limiting current. However, it is important that results obtained from anodic dissolution experiments be confirmed by weight-loss measurements because it is possible that anodic dissolution currents are not equal to corrosion rates in this system. The reason that the anodic dissolution currents could be unrepresentative of free-corrosion rates is that the film formed at the corrosion potential may differ from the film formed at the potential at which the limiting current is measured. For example, hydrogen gas evolution during free corrosion may affect the corrosion rate.[9]

Ellison and Schmeal have used the RCE system to study the corrosion of mild steel in 68% and 78% sulfuric acid at 27°C (81°F), 40°C (104°F), and 60°C (140°F) and over the 0.4 ms^{-1} to 1.6 ms^{-1} velocity range. They found that the corrosion rate was indeed controlled by diffusion of a ferrous ion away from a ferrous sulfate film-liquid interface into the bulk of the acid. They have been able to successfully fit their results to the mass transfer relation applicable to turbulent flow at an RCE and derived earlier by Eisenberg et al.[46] Therefore, it was decided to use the same model to calculate predicted values for the limiting corrosion currents. The expression for the current has already been derived in Section 1.6.2 and found to be:

$$I_L = KzF(c_s - c_b)S \tag{3.1}$$

where K, the mass transfer factor taken from the Eisenberg relation (cf. Section 2.2.2), can be written as:

$$K = 0.079\frac{D}{d}Re^{0.7}Sc^{0.356} \tag{3.2}$$

The limiting current, obtained from the anodic dissolution experiments, was then compared to the calculated current using Equation (3.1). The anodic polarization curve obtained on cast irons showed a well-defined limiting current. As a consequence, the approach applied to mild steel was equally used for the cast irons.

The velocities used in this study ensured that at all times the conditions were turbulent. It has previously been shown (cf. Section 2.2.2) that the RCE operates in the turbulent regime at all but the very low velocities.[19,46] The critical Reynolds number (Re_{CRIT}) value for the transition between laminar and turbulent regimes has been reported to be $Re_{CRIT} \approx 200$.[19] The lowest Reynolds (Re) value used in this study is 2,360. This means, notwithstanding the results of the present investigation, that the Eisenberg relation can be used to calculate the predicted limiting current value.

The Re was calculated using the well-known expression:

$$Re = \frac{dv\rho}{\mu} \qquad (3.3)$$

where d = 31 mm is the diameter of the inner rotating cylinder, i.e., the diameter of the working electrode, and v is the linear velocity of the electrode. Three values were used in this study: 0.81 ms^{-1}, 1.62 ms^{-1}, and 2 ms^{-1} corresponding to rotation speeds of 500, 1,000, and 1,235 rpm, respectively. ρ is the density of the acid in kg/m^3 at the concentrations used and for the temperatures of interest (27°C [81°F] and 40°C [104°F]).[91] They have also been very conveniently represented in a graphical manner by S. Dean as reproduced in Figure 3.5.[12] μ is the viscosity in kg/m/s for which there are a number of convenient sources.[92-94] Again, the data have been collated in a graphical manner by S. Dean (cf. Figure 3.6).[12] The Re in this study varied from 2,360 to 17,873.

The Schmidt number (Sc) is calculated using the formula:

$$Sc = \frac{\mu\rho}{D} \tag{3.4}$$

The viscosity and density are taken from the previous references.[91-94] The diffusion coefficient, D, of ferrous ion in concentrated sulfuric acid has been measured by Ellison and Schmeal with a rotating platinum disc electrode at two concentration levels of 68% and 78%.[9] They proposed an extrapolation formula:

$$D = 1.68.10^{-12}\frac{T}{\mu} \tag{3.5}$$

where T is the temperature in °K and μ is the acid viscosity in centipoise. This formula has been successfully used by S. Dean to calculate corrosion rates in the pipe flow geometry using higher concentrations of acid (i.e., 95%).[12] In the absence of other data, the Ellison and Schmeal formula has been used in this study.

In this way, the value of K can be calculated from Equation (3.2), and the current can be thus predicted from Equation (3.1). For this, the solubility of ferrous sulfate in sulfuric acid, c_s, at the concentrations and temperatures of interest must be known. This is available in the literature, particularly by Kobe, and also in *Solubilities of Inorganic and Organic Compounds* edited by H.L. Silcock.[3,95] Extrapolation and interpolation from the published data are straightforward. Furthermore, the data at 0°, 27°, 40°, and 60°C (32°, 81°, 104°, and 140°F) for the 50 to 100% acid concentration range have been graphically collated by S. Dean and reproduced in Figure 3.7.[12]

Since the runs for the different alloys, acid concentrations, temperatures, and velocities were performed over a length of time, it was necessary to periodically remove a sample of acid for determination of acid ferrous ion concentration, c_b. The method adopted for this measurement has been the orthophenanthroline technique. The method is the subject of an ASTM standard, a summary of which is reproduced in Appendix C.[96] The photometric reading is first converted to an iron percentage, which is then converted to a concentration in moles/cm^3 using the acid density at the concentration and temperature of interest (cf. Appendix C).

The value of the current calculated using Equation (3.1) was reported along with the value of the limiting current measured at the time the acid sample was taken. Each run was reported in tabulated form, and all the data have been arranged for ease of reference in Appendix D. A superscript was used to differentiate between the measured values, i.e., $K^{(1)}$, $CR^{(1)}$, $Sh^{(1)}$ and the calculated values $K^{(2)}$, $CR^{(2)}$, $Sh^{(2)}$. The properties reported that are essential to the subsequent analysis of the data are: alloy type, acid concentration, temperature, velocity, viscosity, acid density, ferrous ion diffusion coefficient, ferrous sulfate solubility, bulk acid ferrous ion concentration, Re, Sc, and measured and calculated mass transfer factors.

The conversion between the measured current values in mA reported in the tables and the corrosion rate values in mm/y is conducted according to the following formula:[97]

$$CR(mm \cdot y^{-1}) = 11.5905\ i$$

where i is the current density in mA/cm^2. Since the surface area of all samples used is 9.74 cm^2, the conversion formula reduces to:

$$CR(mm \cdot y^{-1}) = 1.19\ I$$

where I is the current in mA.

The original derivation is based on:[97]

- the molecular weight of Fe: $55.85\ g \cdot mole^{-1}$
- the density of iron: $7.88\ g \cdot cm^{-3}$
- the valency change of n = 2 corresponding to the reaction: $Fe \rightarrow Fe^{++} + 2e^{-}$

During the experiments in 93% and 98% acid, the measured values for the currents were much higher than the predicted values. The reported Re exponents (0.78 and 0.81) are the values found to fit the data best (cf. Section 4.2.1).

Direct weight-loss measurements were conducted to confirm that the anodic dissolution currents are truly representative of the corrosion currents. The direct weight-loss values were found to be in reasonable accord with the weight loss calculated by integration of the measured current over the experiment duration using the expression:

$$W = \frac{M\Sigma It}{zF} \tag{3.6}$$

where ΣIt is the charge (Coulombs) represented graphically by the surface area of the current-exposure time curve. The calculation is straightforward and the results are reported for some of the systems studied in Table 3.2. Examination of the data shows that the limiting current value measured on the anodic polarization is also equal to the free-corrosion rate.

The direct weight loss is generally higher than the calculated, confirming the difficulty of measuring the high currents experienced at short exposure times.

Table 3.1

Chemical Analysis of the Cast Irons Studied

Elements Determination %	Ductile Iron	Grey Cast Iron	Mondi A	Mondi B
C	3.32	2.99	3.48	3.44
Si	2.80	3.01	3.01	3.00
P	0.019	0.018	0.018	0.019
S	0.016	0.024	0.022	0.012
Mn	0.47	0.92	0.52	0.50
Ni	0.088	0.086	0.28	0.19
Cr	0.095	0.084	0.090	0.088
Mo	<0.05	<0.05	<0.05	<0.05
Cu	0.066	0.064	0.14	0.14
Mg	0.042	0.006	0.032	0.028
B	<0.001	<0.001	0.007	0.003
Ti	<0.01	<0.01	<0.01	<0.01
V	<0.01	<0.01	<0.01	<0.01
Al	0.007	<0.005	0.006	0.006
Sn	<0.01	<0.01	<0.01	<0.01

Table 3.2

Calculated and Measured Weight Loss for Some of the Conditions Tested. The Two Values Correspond to the Two Working Electrodes Used: WE 1 and WE 2, Respectively

Alloy	Acid Concentration %	T °C	Direct Weight Loss (g) W1	Calculated Weight Loss (g) W2	Deviation % (W1-W2)/W1
Ductile Iron	98	40	0.28759 0.30739	0.25356 0.25356	+11 +17
Mondi A	98	40	0.20449 0.21444	0.21940 0.21940	-7 -2
Grey Cast Iron	98	40	0.22142 0.18454	0.20464 0.20464	+7 -11
Mild Steel	93	40	0.45813 0.47149	0.40670 0.40670	+11 +13
Ductile Iron	98	27	0.09264 0.08326	0.07596 0.06738	+18 +19
Mondi A	98	27	0.15828 0.15789	0.13597 0.13231	+14 +16
Mondi B	98	27	0.15640 0.15756	0.14743 0.14743	+6 +6
Grey Cast Iron	98	27	0.07317 0.09034	0.06221 0.07451	+15 +17

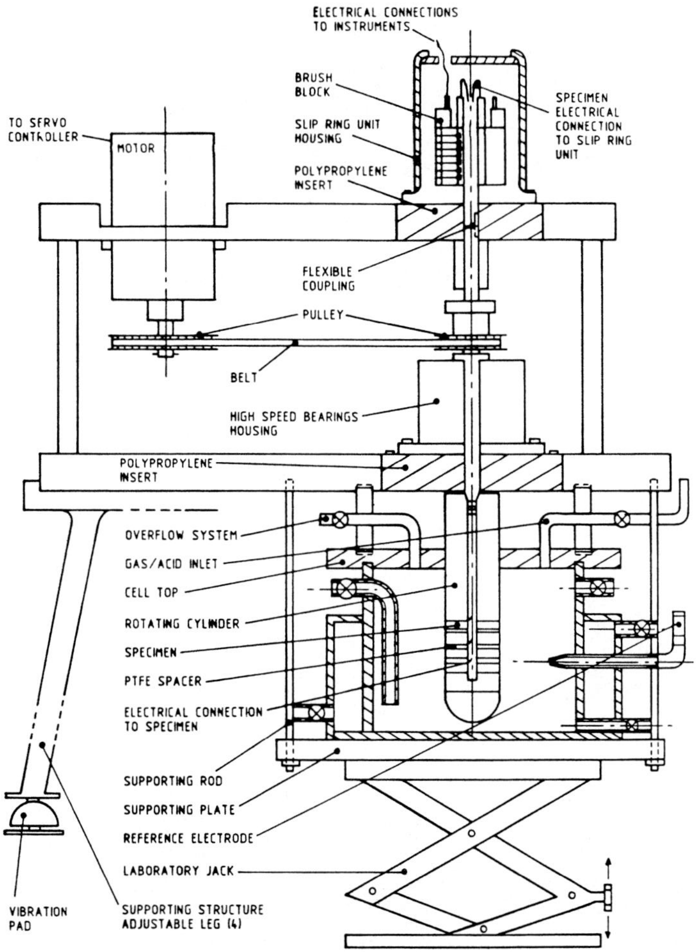

Figure 3.1 Schematic diagram of the rotating cylinder assembly.

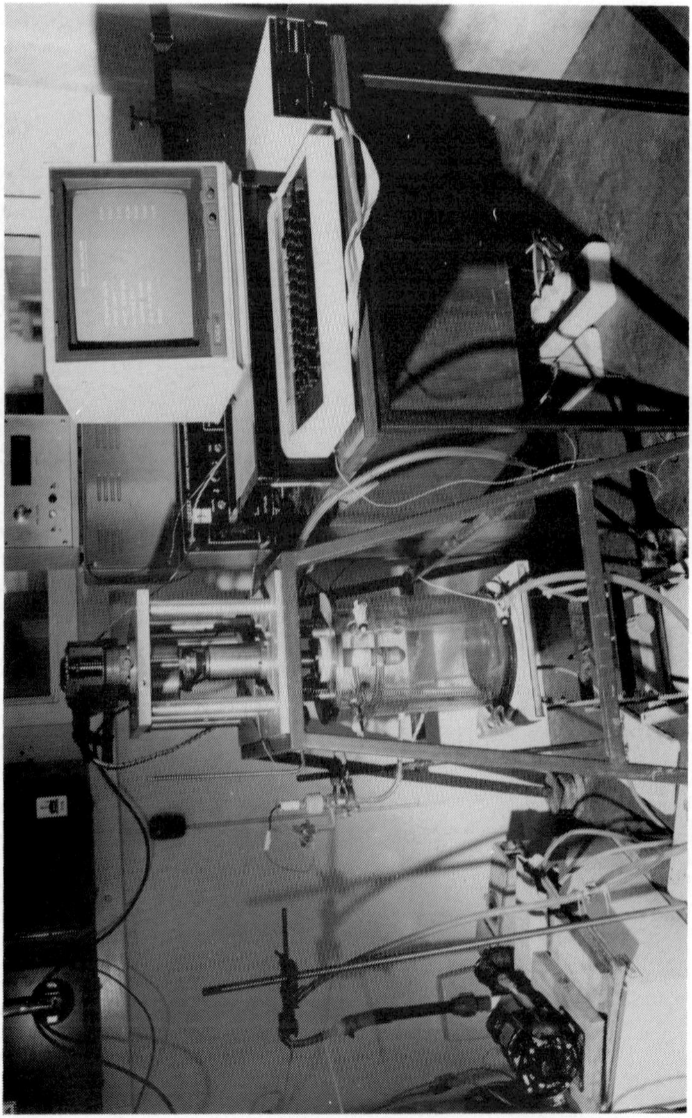

Figure 3.2 Photograph of the rotating cylinder electrode assembly.

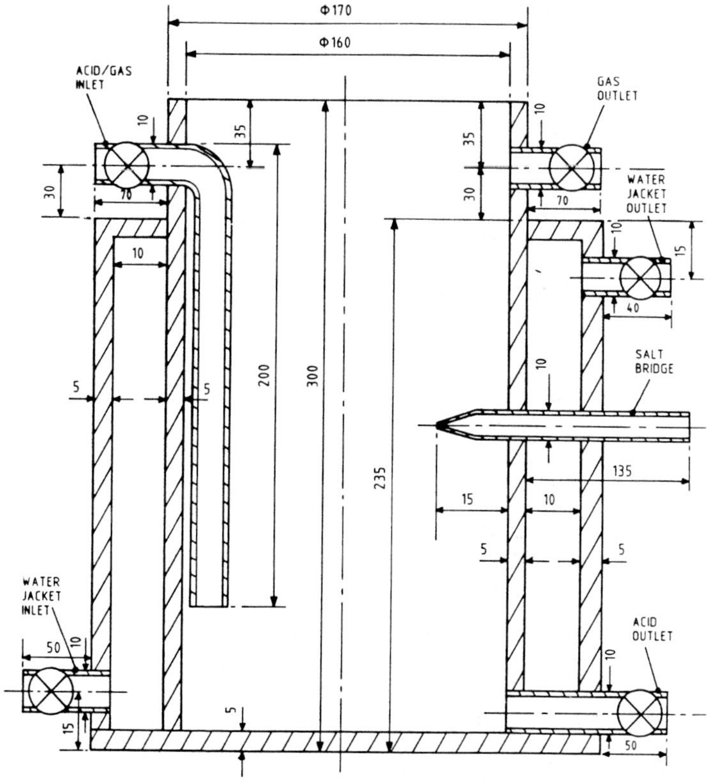

Figure 3.3 Schematic diagram of the rotating electrode cell—all dimensions in millimeters. Sketch not to scale.

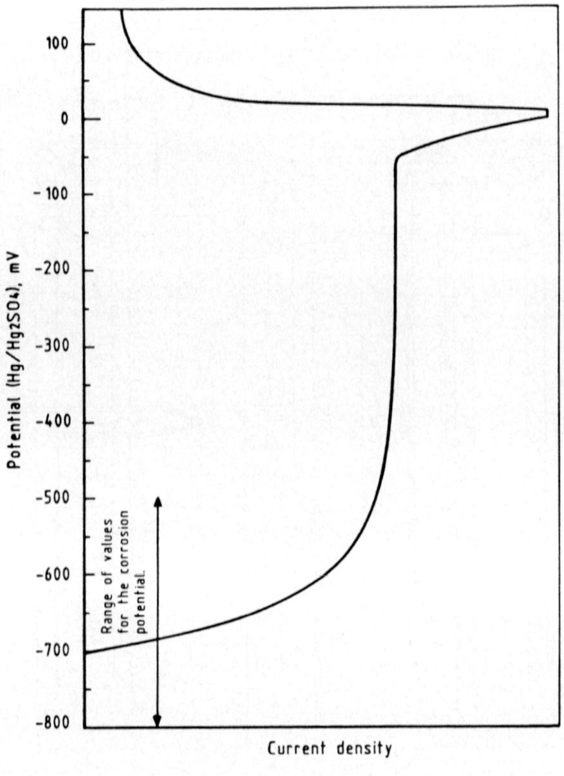

Figure 3.4 The anodic polarization curve of mild steel and cast irons in concentrated sulfuric acid showing the essential features observed.

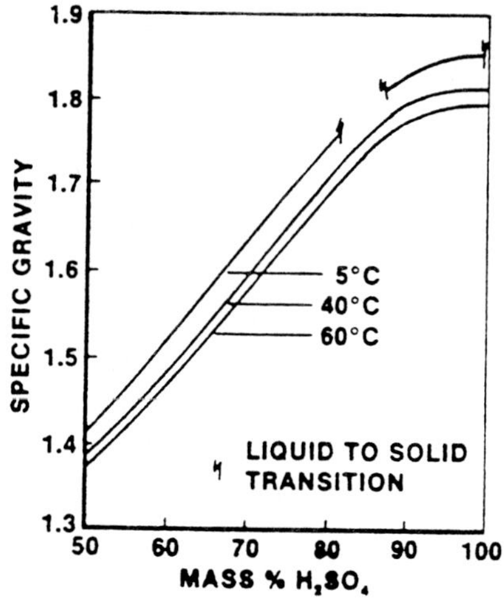

Figure 3.5 Specific gravity of concentrated H_2SO_4.[12] © Air Products and Chemicals, Inc. Reprinted with permission of Air Products and Chemicals, Inc.

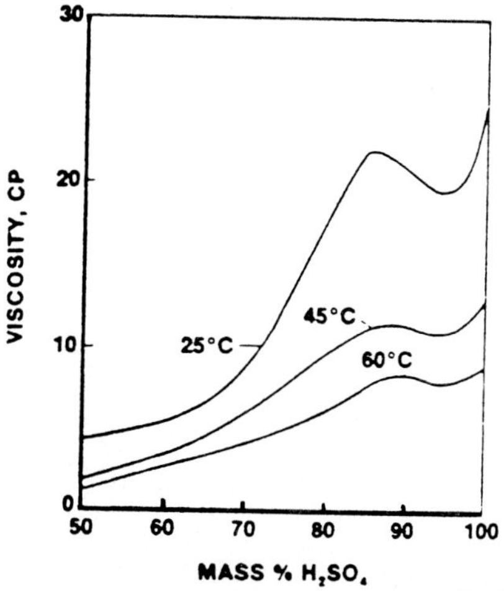

Figure 3.6 Viscosity of concentrated H_2SO_4.[12] © Air Products and Chemicals, Inc. Reprinted with permission of Air Products and Chemicals, Inc.

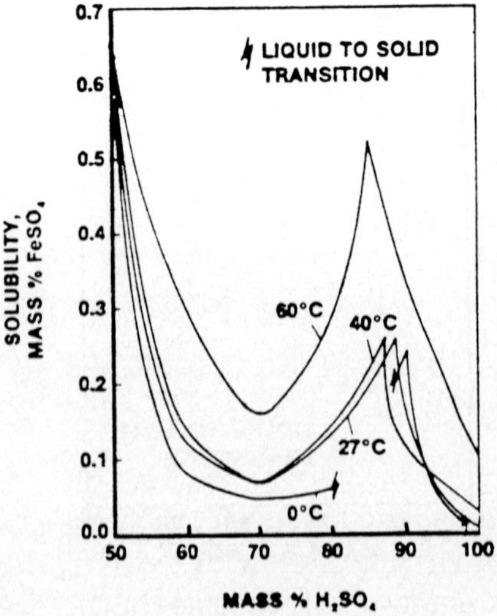

Figure 3.7 Solubility of $FeSO_4$ in concentrated H_2SO_4.[12] © Air Products and Chemicals, Inc. Reprinted with permission of Air Products and Chemicals, Inc.

CHAPTER
FOUR

EXPERIMENTAL RESULTS

4.1 INTRODUCTION

In the previous chapter, the rotating cylinder was described together with the instrumentation that constitutes the overall experimental set-up. The experimental techniques have been detailed, together with the microstructural composition of the cast irons under investigation.

The rotating cylinder electrode (RCE) apparatus has been used to carry out an extensive testing program of the following alloys:

Mild Steel
Ductile Iron
Mondi A
Grey Cast Iron
Mondi B

The linear velocities employed were in the range 0.81 to 2 ms^{-1}, corresponding to rotating speeds of 500 rpm to 1,235 rpm for the electrode. These velocities are typical, or somewhat above, the values used and considered acceptable for cast iron and steel equipment handling concentrated acid in industry.[2,5]

The acid concentrations tested were 98% and 93%. Analar sulfuric acid (analytical grade) was used exclusively for the present work.

Two temperatures have been used throughout; namely, 27°C (81°F) and 40°C (104°F). These are typical of the limits at which these types of alloys are expected to be used in the design of equipment such as pipework and storage tanks destined for the handling of concentrated sulfuric acid.[2,5,12] In addition, it was decided to limit the temperature to

40°C (104°F) because the increase in acid fumes with temperature can lead to premature failure of the high-speed bearings in the rotating cylinder apparatus.

For each alloy tested throughout this study, two identical samples were mounted together on the rotary shaft and used as separate working electrodes. In addition to providing a back-up sample, this also gave an indication of the consistency of the results.

The literature survey showed the existence of a single reference where experimental data on mild steel in 68% sulfuric acid at 27°C (81°F) and 40°C (104°F) was obtained using the rotating cylinder method.[9] The published results were correlated using the mass transfer relation derived by Eisenberg et al. (cf. Section 2.2.2).[46] Therefore, it was decided to check the RCE apparatus against the published data and to compare the original findings obtained in this study to the same model. This will be referred to throughout this study as the *Eisenberg relation* or the *model*.

The present work concentrated on investigating the experimental value of the Reynolds (Re) exponent for the systems and the range of Re and Schmidt (Sc) numbers used.

The values of the constant, 0.079, and the Sc exponent, 0.356, found by Eisenberg and confirmed by a number of workers were assumed to be valid for all of the data generated in this study.[9,13]

4.2 STEEL IN CONCENTRATED SULFURIC ACID

4.2.1 Introduction

Corrosion of mild steel in concentrated sulfuric acid has been reported to be controlled by the convective diffusion rate of the ferrous ion away from the corroding surface.[9-12] This diffusion model assumes that the surface is covered by a layer of ferrous sulfate at its saturation concentration level, c_s. Since the bulk of the acid will have a concentration of ferrous ion, c_b, a concentration driving force, $c_s - c_b$, will be established across the diffusion boundary layer of thickness, δ_d.

As fluid flow or electrode rotation speed increases, thinning of this layer will ensue, and the corrosion current, which is also the measured limiting current, will increase. Furthermore, as the bulk concentration, c_b, increases, the concentration driving force decreases, and the corrosion current given by:

$$I_{corr} = I_L = \frac{zFD(c_s - c_b)S}{\delta_d}$$

will decrease. Those aspects of the corrosion predicted by the model will now be investigated in the light of experimental results.

The results of Ellison and Schmeal, who studied a number of steels of different compositions, also proved an earlier observation by Damon that steel composition, particularly carbon and copper content, is a minor variable in determining the general corrosion resistance of carbon steel in concentrated acid.[9,11] For acid strengths below approximately 50%, corrosion rates are strongly influenced by the carbon content of the steel.[11] As a consequence, it was decided to carry out the present study on a single steel composition (cf. Section 3.2).

S. Dean has analyzed experimental results in the literature and found that the deviation of measured data from the correlation proposed by Eisenberg is strongly dependent upon the investigator and that this can vary from less than 10% to more than a factor of 2.[9,10,12,46]

Moreover, Ellison and Schmeal found the limit that can be expected on the accuracy of the predictions of the model to be about $\pm 25\%$.[9] The reason for this is that the cumulative errors in the measurement of diffusion coefficient, ferrous ion solubility, ferrous ion concentration in the bulk acid, and corrosion rate all influence the agreement between the measured corrosion rate and the predictions of the model.

All the experimental results obtained on mild steel in this study showed that the relative difference between measured and calculated current varied with time. Therefore, to obtain a plot of $Sh/Sc^{0.356}$ vs Re, the parameter, $Sh/Sc^{0.356}$, was calculated from the measured corrosion current averaged over the duration of each run at a given velocity. In this way, a single value, and thus a single point on the graph, was obtained for each value of Re. The best fit line through the data points was obtained by plotting log $Sh/Sc^{0.356}$ against log Re for all the data points with an additional point at (-1.1,0) corresponding to the assumed constant 0.079. The value of the slope will be considered in detail in the next sections, but it was found to vary between 0.7 and 0.81 for the acid concentrations, temperature, and velocity range used.

The presumed convective diffusion corrosion mechanism describes the uniform dissolution of the corroding surface (cf. Section 2.4.2). Therefore, it was necessary to examine the corroded surfaces at the end of each run. These were observed to have remained smooth and shiny, indicating, in effect, uniform corrosion as expected.

Corrosion potential measurements indicated potential values in 98% sulfuric acid at 27°C (81°F) between -730 mV and -690 mV for the 0.81 to 2 ms^{-1} velocity range. Similar values were found at 40°C (104°F). In 93% sulfuric acid at 27°C (81°F), the values were between -820 mV and -800 mV for the 0.81 to 2 ms^{-1} velocity range. Similar values were also found at 40°C (104°F). The potentials were able to stabilize at the above values after the first few hours of immersion. All corrosion potentials reported in the present work are measured with respect to an Hg/Hg_2SO_4 reference electrode. Figure 4.1 shows the measured variation of corrosion potential with time for mild-steel specimens.

4.2.2 Mild Steel in 68% Sulfuric Acid

The objective of this experiment was twofold. First, it was important to check that the RCE assembly and measurement systems were providing valid data. As detailed below, Ellison and Schmeal have published data for mild steel in 68% sulfuric acid, and their results could be directly compared to the results of this study to give confidence in the reliability of the data in 98% and 93% sulfuric acid.

Secondly, the data would provide additional baseline information for comparison with the test program. The RCE technique has been used by

Ellison and Schmeal on mild steel in 68% sulfuric acid at 27°C (81°F), 40°C (104°F), and 60°C (140°F) and for a velocity range of 0.4 to 1.6 ms^{-1}.[9] They confirmed that corrosion is limited by the rate of convective mass transfer of ferrous ions from a ferrous sulfate film-liquid interface into bulk acid. The corrosion currents were found to vary with velocity to the nth power where n = 0.7.

The present investigation showed the results on mild steel in 68% sulfuric acid at 27°C (81°F) and 40°C (104°F) over the 0.81 to 2 ms^{-1} velocity range to be in agreement with those of Ellison and Schmeal. The full set of results is shown in Appendix D, Tables D.1 through D.5.

The apparatus used by Ellison and Schmeal had an outer concentric platinum mesh auxiliary electrode and an overflow system so that no bubbles or vortex formation was allowed to occur.[9] However, for the present study (cf. Section 3.1), an AISI 316 (UNS S31600) stainless steel auxiliary electrode mounted on the rotating shaft was used. Substantial practical difficulties were encountered when attempts to seal the cell were made; therefore, some vortex shedding was present at the highest velocity. To avoid the presence of oxygen in the acid, a nitrogen blanket was used for all tests.

These variations in design turned out to be of little consequence to the mass transfer mechanism since our results faithfully reproduced theirs. Therefore, the particular geometry of the present study could be added to the number of geometrical variations investigated by Eisenberg and were found to have a small influence on the currents obtained and, hence, on the mass transfer correlation (cf. Section 2.2.2).[46]

The best-fit line (Figure 4.2) through the data points at 40°C (104°F) had a slope of 0.7 with a maximum deviation of ±20% around the model prediction. This slope was 0.675 (Figure 4.3) for the data points at 27°C (81°F). The value of 0.7, quoted by Eisenberg et al., is said to fit data obtained for five different electrochemical systems and over a vast range of Re and Sc (cf. Section 2.2.2).[46] Therefore, the present results confirm those of Ellison and Schmeal and are an excellent check for the apparatus giving added confidence to the results of this study.[9]

4.2.3 Mild Steel in 98% Sulfuric Acid at 27°C (81°F)

As far as the author is aware, this system has not been studied and reported in the open literature. The results should make an interesting comparison with those at 68% sulfuric acid.

Mild steel in 98% sulfuric acid at 27°C (81°F) and over the 0.81 to 2 ms^{-1} velocity range showed a behavior characteristic of corrosion controlled by the rate of convective mass transfer of ferrous ion from a ferrous sulfate saturated layer; the corrosion currents were observed to decrease with the bulk acid ferrous ion concentration and to increase with flow.

The best-fit line through the data points obtained had a slope of 0.77 (Figure 4.3). The maximum deviation around this experimentally derived correlation:

$$Sh = 0.079\ Re^{0.77}\ Sc^{0.356}$$

was +29%. The complete set of results is shown in Appendix D, Tables D.6 through D.8.

A number of workers have predicted the dependence of the mass transfer rate on velocity raised to the power of 0.7 to be true for the 50 to 99% acid concentration range.[9,12,76] This generalization assumes the influence of the electrochemical system itself (i.e., acid concentration) and the effect of the range of Re and Sc numbers used to be of little importance. However, the value of 0.77 found at the 98% concentration level is significantly different from the value of 0.7 found at 68% and suggests that the generalization is invalid at high acid strengths.

4.2.4 Mild Steel in 98% Sulfuric Acid at 40°C (104°F)

An increase in temperature from 27°C (81°F) to 40°C (104°F), for the same acid concentration, leads to changes in viscosity, diffusion coefficient, and ferrous sulfate solubility, and thus to an increase in the Re and a decrease in the Sc. The increase of the ferrous sulfate solubility, together with the changes in the other parameters, was found to lead to a dramatic increase in current. At 0.8 ms^{-1}, and for a bulk acid ferrous ion concentration of $c_b = 0$, this was found to be increased by a factor of 2.95. The observed behavior was characteristic of a corrosion mechanism controlled by the rate of convective mass transport of the ferrous ion as already observed at 27°C (81°F).

The best-fit line (Figure 4.2) through the data points showed a slope of 0.81, the maximum deviation around the experimental line being

+28%. The complete set of results is listed in Appendix D, Tables D.9 through D.11.

4.2.5 Mild Steel in 93% Sulfuric Acid at 27°C (81°F)

Measurements on mild steel in 93% sulfuric acid at 27°C (81°F) in the 0.81 to 2 ms^{-1} velocity range showed the characteristic behavior of corrosion controlled by the rate of convective mass transfer of ferrous ion from a ferrous sulfate saturated layer. While this is similar to the results given above, the major difference between 98% sulfuric acid and 93% sulfuric acid at 27°C (81°F) is a substantial increase in the ferrous sulfate solubility, i.e., from $2.288 \cdot 10^{-6}$ moles/cm^3 to $10.816 \cdot 10^{-6}$ moles/cm^3.

This increase is reflected by a higher corrosion rate in 93% sulfuric acid. The complete set of results is shown in Appendix D, Tables D.12 through D.14. The rather modest changes in viscosity, diffusion coefficient, and acid density led to a relatively small increase in Re and Sc numbers.

The best-fit line (Figure 4.3) through the data points obtained on mild steel in 93% sulfuric acid at 27°C (81°F) had a slope of 0.77 with a maximum deviation around the observed line of -15%.

In conclusion, the influence of an acid concentration change from 98% to 93% at 27°C (81°F) and for the 0.81 to 2 ms^{-1} velocity range on the Re number exponent is very small. The dominant effect is in the respective ferrous sulfate solubilities and the consequent higher corrosion rates.

4.2.6 Mild Steel in 93% Sulfuric Acid at 40°C (104°F)

An increase in temperature from 27°C (81°F) to 40°C (104°F) in 93% sulfuric acid leads to a significant change in viscosity and diffusion coefficient and thus, higher Re and lower Sc numbers. The corrosion is again characteristic of corrosion controlled by the rate of convective mass transfer of the ferrous ion as previously observed throughout these series of tests on mild steel.

The corrosion currents were observed to increase by about a factor of 2. This is the result of the combined effect of the higher Re numbers and diffusion coefficient. However, there is also a contribution from an increase in the velocity exponent. Appendix D, Tables D.15 through D.17 summarize the full set of data.

The best-fit line (Figure 4.2) through the data points obtained on mild steel in 93% sulfuric acid at 40°C (104°F) has a slope of 0.81 with a maximum deviation around the experimentally derived correlation of +17%.

4.2.7 Concluding Remarks on Corrosion of Mild Steel

The mass transfer relation derived by Eisenberg and confirmed by Schmeal in 68% sulfuric acid has been investigated as a possible model for corrosion of steel in 98% and 93% sulfuric acid.[9] A mechanism whereby corrosion is completely controlled by diffusion has been confirmed.

In the work of Eisenberg et al., the velocity exponent was higher for the dissolution experiments (e.g., benzoic acid dissolution into water) than for the redox experiments (e.g., cathodic reduction of ferricyanide). A least squares analysis of their results was performed by Makrides et al., who reported the velocity exponent in the case of roughening cylinders during dissolution must be higher, and it also depended on the Re number range.[60] He found the value to be 0.68 for the interval Re = 1,000 to 10,000, while the value was 0.77 for Re = 30,000 to 100,000.

However, the variations observed in the present study are due to the different acid concentrations (68%, 98%, and 93% sulfuric acid) and temperatures used since no surface roughening of the corroded surfaces could be detected. No mechanistic insight can so far be gained from the variations reported apart from stating that the higher values found must be attributed to an effect of the dissolution mechanism on the hydrodynamic conditions under which mass transfer occurs. This has also been the conclusion of Makrides et al.[60] For their part, Eisenberg et al. did not attempt any explanation of the 0.7 value found other than reporting its applicability to the systems studied over the range of Re and Sc numbers of interest.[46] These fundamental points will be examined further in the discussion section.

4.3 CAST IRONS IN CONCENTRATED SULFURIC ACID AT 27°C (81°F)

4.3.1 Introduction

The results described below were obtained for a series of four different cast irons tested in 98% sulfuric acid at 27°C (81°F) over a 0.81 to 2 ms^{-1} velocity range. Figure 4.3 shows all the collected data plotted as $Sh/Sc^{0.356}$ vs Re on a log-log scale.

When freshly prepared surfaces of cast irons first make contact with the acid, they show anomalously high-corrosion currents compared with the values predicted by the model. It should also be noted that the corrosion potentials were at that time very active. These high currents were observed to be strongly dependent on flow, implying that a mass transfer mechanism was operative.

Following immersion, the current densities decreased and the potentials steadily shifted to more noble values. The duration of this period of current decay was dependent on the type of cast iron used and the rotation speed to which it was first subjected.

After a period of time, the corrosion currents fell below the theoretical values. This was also accompanied by a reduction in their flow sensitivity, which was observed to vary with the cast iron used, rotation speed, and time. The corrosion potentials were also observed to be dependent upon the same parameters.

Examination of the corroded cast iron surfaces at the conclusion of each run revealed a roughened, grainy film. This was not observed on any of the exposed mild-steel surfaces that kept the as-prepared smooth and shiny aspect suggestive of uniform corrosion. Cast irons contain in their microstructure a number of, more or less, corrosion-resistant morphological phases largely absent from the microstructure of mild steel. As a result, the corrosion of cast iron alloys is not completely uniform, and a film in the form of a graphitic residue and richer in the noble constituents, such as graphite, silicon, phosphide eutectic, and, to a lesser extent, carbides, is left behind.[98]

The combined effect of the original cast iron composition and morphology, the extent of corrosion, the flow conditions, and exposure time will lead to graphitic residues different in composition, morphology, flow resistance, and thickness. In turn, these will gradually lead to a stifling of the corrosion rate.

4.3.2 Mondi A in 98% Sulfuric Acid at 27°C (81°F)

This alloy was first subjected to a rotation speed of 1,000 rpm, corresponding to a linear velocity of 1.62 ms^{-1}. When the samples made contact with the acid, the first corrosion potential observed was -800 mV. The measured corrosion current was initially high, but within approximately 12 h dropped to a value comparable to that theoretically calculated. Figure 4.4 shows the variation of the calculated and measured currents with time and rotation speed. The full set of results is shown in Appendix D, Tables D.18 through D.20. The measured variation of corrosion potential with time and velocity is shown in Figure 4.5.

At the end of the first run of 98 h at 1,000 rpm, the measured corrosion current was only 7% below that theoretically calculated; the corrosion potential on both electrodes used was around -650 mV.

Following this first test, the rotation speed was reduced to 500 rpm corresponding to a linear velocity of 0.81 ms^{-1} The system was kept at that speed for 24 h. The corrosion potentials stayed within the range of values given above, and the measured current was only marginally different from that theoretically calculated.

Finally, the rotation speed was increased to 1,235 rpm corresponding to a linear velocity of 2 ms^{-1}. The measured corrosion current was, after 24 h, consistently below the theoretically calculated value by about 20%. At the end of the run, the current on one of the electrodes dropped to about 33% below the theoretical value. Its corrosion potential was then -560 mV, while the potential was unchanged for the other electrode. It appears that there is a difference in the films formed on the two electrodes used. One was able to reduce the current further and shift the potential to more noble values than the other. However, considering that the limit on the accuracy of the predictions of the model was about ±20%, and except for the corrosion current measured on one of the electrodes at the end of its exposure at 1,235 rpm, the behavior observed for Mondi A in 98% sulfuric acid at 27°C (81°F) could be considered similar to that of mild steel for the same conditions.

Examination of the corroded surfaces at the conclusion of the tests did not show any noticeable roughening. This was also the case for the steel surfaces tested under similar conditions and is consistent with the measured corrosion current behavior.

4.3.3 Ductile Iron in 98% Sulfuric Acid at 27°C (81°F)

This alloy was first tested at a linear velocity of 2 ms^{-1}. The corrosion currents were initially high and fell very quickly after 10 h to reach values close to those theoretically calculated. Figure 4.6 shows the variation of the calculated and measured currents with time and rotation speed. The full set of results is listed in Appendix D, Tables D.21 through D.23. It has been observed that the current decay transient is shorter at higher velocities. Following the transient, a distinct difference in current between the two otherwise identical electrodes used was observed. Eventually, after 51 h at 1,235 rpm, the two electrodes showed currents lying at 34% and 15% below the theoretical value, respectively. Upon contact with the acid, the corrosion potential values were around -800 mV. These values became noble very quickly and finally stabilized at around -690 mV after 51 h at 1,235 rpm. The measured variation of corrosion potential with time and velocity is shown in Figure 4.5.

The rotation speed was then lowered to 1,000 rpm. The difference in currents between the two electrodes started to decrease until, after 28 h at 1,000 rpm, both currents were found to be at about 30% below the theoretical value. The corrosion potentials varied a little to later stabilize at around -700 mV.

Finally, when the rotation speed was further decreased to 500 rpm, the currents on both electrodes were found to be 10% below the calculated value with their corrosion potentials at about -680 mV. Examination of the corroded surfaces revealed a dull appearance with no appreciable increase in roughness. This is consistent with the small current decrease below the theoretical value and the corrosion potential values of between -680 mV and -700 mV. These values were around -700 mV (Figure 4.1) for mild steel in 98% sulfuric acid at 27°C (81°F) where uniform corrosion and no surface film was observed.

4.3.4 Grey Cast Iron in 98% Sulfuric Acid at 27°C (81°F)

This alloy was first subjected to a rotating speed of 1,235 rpm. Upon contact with the acid, the corrosion potentials were at -840 mV and increased very quickly to around -700 mV. Figure 4.5 shows the measured variation of corrosion potential with time and rotation speed. The corrosion currents measured were initially high and reached, in the first 24 h, values comparable to those theoretically calculated. Figure 4.7 shows the variations of the measured and calculated currents with time and rotation speed, while Tables D.24 through D.26 of Appendix D list the complete set of results.

After this initial period, a current difference began to develop between the two electrodes used; the respective current values were 58% and 32% below that theoretically calculated. The corrosion potentials were then around -650 mV.

Following the first test condition, the rotation speed was decreased to 1,000 rpm. The corrosion potential values stabilized at -640 mV, while the same two electrodes still showed a large difference in their respective currents at 51% and 30% below the theoretical value but slightly less than at 1,235 rpm.

Finally, when the rotation speed was decreased to 500 rpm, the corrosion potential values were only marginally more noble at -620 mV. The measured currents were then at 34% and 16% below the theoretical values, respectively.

This decrease in corrosion current was accompanied by a reduction in flow sensitivity that was higher for the low-current electrode. Lowering the rotation velocity from 1,235 rpm to 1,000 rpm led to virtually no change in its measured corrosion current, whereas the calculated currents predicted a decrease of around 14%. For the next lower rotation speed of 500 rpm, the observed corrosion current decrease was only 16%, while the predicted value was 39%. Meanwhile, the same measured values for the other electrode, at respectively 16% and 32%, were comparable to those theoretically calculated.

The following aspects of grey cast iron corrosion resistance in 98% sulfuric acid at 27°C (81°F) are notable:

- A significant decrease in corrosion current and a partial loss of flow sensitivity occurs for this system. Both phenomena were more pronounced at the higher velocities.
- There is a considerable noble shift in corrosion potentials from about -800 mV to around -620 mV.
- Examination of the used specimens at the end of the tests revealed a noticeable roughening of the corroded surfaces.
- The extent of the above observations was less pronounced on either Mondi A or ductile iron under similar conditions.

4.3.5 Mondi B in 98% Sulfuric Acid at 27°C (81°F)

This alloy was first tested at the 500 rpm rotating speed. The corrosion currents observed were initially very high and decreased very slowly to reach, after 88 h through the run, values comparable to that theoretically calculated. This confirms a previous observation that the higher the initial velocity, the faster the currents will decrease to or below the theoretical value. Figure 4.8 shows the variations of the measured and calculated corrosion currents with time and rotation speed. A complete listing of results is shown in Appendix D, Tables D.27 through D.29. The corrosion potential values were then around -700 mV. Figure 4.5 shows the measured variation of corrosion potential with time and velocity.

Following the first test conditions, the rotation speed was increased to 1,000 rpm, the corrosion current continued to decrease to reach, after 77 h, a value of 39% below that theoretically calculated. The corrosion potential values were also observed to become more noble at around -670 mV.

Finally, the rotation speed was increased to 1,235 rpm, and the corrosion current continued decreasing to reach, after 68 h, a value 58% below that theoretically calculated. The corrosion potential values were also observed to become more noble between approximately -650 mV and -620 mV.

This decrease in corrosion current was accompanied by a reduction in the sensitivity of corrosion current to flow. When the rotation speed was increased from 500 rpm to 1,000 rpm, the ratio of the calculated currents predicted an increase of 60%, whereas the measured increase was only 20%.

In conclusion, the decrease in corrosion current and the partial loss of flow sensitivity, which were both higher at the higher velocities, the shift in corrosion potential to more noble values, and a marked change in roughness of the corroded surfaces make the behavior of Mondi B similar to that of grey cast iron.

4.4 CAST IRONS IN 98% SULFURIC ACID AT 40°C (104°F)

4.4.1 Introduction

These series of tests have been conducted for all four cast irons under investigation. The alloys were initially subjected to the low rotation speed of 500 rpm. This was then increased to 1,000 rpm and finally to 1,235 rpm. The results, plotted as $Sh/Sc^{0.356}$ vs Re, are shown in Figure 4.2, while Table D.34 in Appendix D summarizes the measured values of $Sh/Sc^{0.356}$ for all four cast irons tested. Each time a new rotation speed was applied, the corrosion current and potential within the first hour were measured to give their instantaneous variation with flow. The sensitivity of corrosion current to flow was found to be significantly more reduced at 40°C (104°F) than at 27°C (81°F). Moreover, the extent of this effect showed a marked dependence on the alloy type. For example, the grey cast iron and Mondi B showed a lower sensitivity to flow than Mondi A and ductile iron as previously observed at 27°C (81°F). With increasing time, the measured current was found to be a smaller and ever decreasing fraction of the calculated value. This fraction was considerably less than that measured at 27°C (81°F) under similar conditions. It was also found to be smaller for Mondi B and grey cast iron than for Mondi A and ductile iron. Furthermore, the loss of flow sensitivity and the relative decrease in measured corrosion current with time were both found to be higher at higher velocities. With the decrease in corrosion currents as time passed, the corrosion potentials were seen to become increasingly noble. This was particularly so for Mondi B and grey cast iron and was more marked at 40°C (104°F) than 27°C (81°F). At the end of the tests, examination of the corroded surfaces showed a more pronounced grainy and roughened surface morphology than that at 27°C (81°F).

It is almost certain that the observed growth of surface films is responsible for the behavior observed. This will depend upon the alloy, temperature, time, and velocity as does the corrosion potential, corrosion current decrease, and flow sensitivity.

4.4.2 Ductile Iron in 98% Sulfuric Acid at 40°C (104°F)

At the first applied rotation speed of 500 rpm, the corrosion currents measured were initially very high with values at around 0.616 mA/cm^2, approximately three times higher than that theoretically calculated. However, these high-current densities were of a transient nature, and the observed values dropped in succession to 0.297 mA/cm^2 after 10 h, 0.205 mA/cm^2 after 24 h, and 0.184 mA/cm^2 after 32 h. This last value was just above the theoretically calculated value. Figure 4.9 shows the variation of the measured and calculated corrosion currents with time and rotation speed, while the full set of results is listed in Appendix D, Table D.30. However, this observed current decrease was not related to a corresponding high increase in bulk acid iron contamination, c_b, or a corresponding decrease in the concentration driving force, c_s - c_b (cf. Appendix D, Table D.30). Since a diffusion-controlled corrosion current is proportional to the product of c_s - c_b by the mass transfer factor, K, this means that there must now be an added resistance to convective mass transfer leading to a lower mass transfer factor than that predicted by the simple diffusion model. During the run at 500 rpm, the measured corrosion current showed a decrease by a factor of 3.72, whereas the calculated current predicted a much smaller decrease, i.e., 1.20.

A point worth noting here is that the variation in measured current occurs in two stages (see Figure 4.9); an observation that was found to be valid within experimental variations for all cast irons tested and more pronounced at 40°C (104°F) than at 27°C (81°F). In the first period, the current density decreases rapidly to a value comparable to that theoretically calculated. In the second period, the current decreases more slowly to below that theoretically calculated. For the first period of approximately 30 h, when the currents were high, they were also observed to vary with flow. As the measured currents approached and fell below that theoretically calculated, it was noticed that there was a distinct reduction in their flow sensitivity. After 59 h at the end of the 500 rpm run and the beginning of the next at 1,000 rpm after 1 h, the ratio of the calculated currents predicted a relative increase of 1.57, whereas the corresponding ratio of the measured currents was only 1.14. Once dampening of the corrosion current flow sensitivity has been established, it was observed to increase with time as the ratio of measured-to-calculated current continued to decrease. It was also observed to increase with velocity as the next higher rotation speed of 1,235 rpm led to no appreciable rise in measured current.

At the end of the tests, the corroded surfaces showed a roughened morphology with the existence of a surface film, as revealed by visual examination and light microscopy. Before any further examination of the surfaces or treatment of the electrochemical data, it can be stated that with the passage of time, corrosion was more and more controlled by diffusion through the surface film. This was at the expense of corrosion control by the presumed convective diffusion rate of ferrous ion from a ferrous sulfate saturated film.

Corrosion potential measurements indicated potential values for both electrodes used of around -820 mV upon contact with the acid. Figure 4.10 shows the measured variation of corrosion potential with time and rotation speed. These values then increased to around -740 mV within 15 minutes, showed some instability through the run, and finished around -710 mV after 59 h at 500 rpm. As the rotation speed was further increased to 1,000 rpm, the corrosion potentials measured did not change appreciably, stabilizing at around -705 mV. Finally, at 1,235 rpm, they reached slightly different values at around -690 mV.

4.4.3 Mondi A in 98% Sulfuric Acid at 40°C (104°F)

This alloy was subjected to the same conditions as ductile iron. The rotation speeds were applied in the same sequence: low to high. The length of time each alloy was held at any particular velocity was similar. Therefore, the results are directly comparable.

The corrosion current densities measured were initially very high, 0.626 mA/cm^2 after 1 h. Figure 4.11 shows the variation of the measured and calculated currents with time, while Appendix D, Table D.31, gives the full set of results. These large current densities were of a transient nature and were later observed to decrease at a slower rate and at values lower than that theoretically calculated. This behavior is very similar to that exhibited by ductile iron at 40°C (104°F) and described in the previous section (cf. Appendix D, Tables D.30 and D.31 for a comparison).

Following this first test, the rotation speed was increased to 1,000 rpm. As a result, the measured current increased by a factor of only 1.16, whereas the calculated currents ratio predicted an increase by 1.62. The difference in these two ratios is the consequence of the reduced role with time of the presumed convective diffusion control. This attenuating effect, or loss of flow sensitivity, was observed to increase with time and rotation speed in a manner similar to that just described for ductile iron.

Corrosion potential measurements indicated potential values at around -790 mV upon contact with the acid. Figure 4.10 shows the measured variation of corrosion potential with time and rotation speed. These values then started to become more noble very quickly. They reached -725 mV after 45 minutes and continued to increase steadily with time. At the end of the 61 h of the 500 rpm run, there was a difference in potential between the two otherwise identical electrodes used at -700 mV and -685 mV, respectively. Upon the increase in rotation speed to 1,000 rpm and through the next 25 h of the run, the potential of the first electrode did not change appreciably at -690 mV. Meanwhile, it increased to -655 mV for the other electrode. With the final increase in rotation speed to 1,235 rpm, the potentials continued their slow shift to more noble values to reach -670 mV and -615 mV, respectively, for the two electrodes used.

4.4.4 Grey Cast Iron in 98% Sulfuric Acid at 40°C (104°F)

This alloy was tested under the same conditions as the two previous alloys, ductile iron and Mondi A. The corrosion current densities observed were initially very high: 0.585 mA/cm^2 at 1 to 2 h. Figure 4.12 shows the variation of the measured and calculated currents with time and velocity, while Table D.32 in Appendix D gives the complete set of results. These currents were of a transient nature and were later observed to decrease at a lower rate and at values below that predicted by the model.

Following this first test, the rotation speed was increased to 1,000 rpm, and whereas the calculated currents predicted an increase of 62%, the measured current rise was only 7%. This corrosion current increase is lower than those measured in similar conditions for Mondi A at 16% and ductile iron at 14%, confirming an earlier observation at 27°C

(81°F) that the effectiveness of the surface film to dampen convective diffusion control is strongly dependent on the alloy.

During these tests (cf. the c_b column in Appendix D, Table D.32), the acid was replenished after 24 h at 1,000 rpm. This should have led to a substantial increase in current; however, a rather modest increase was observed to confirm further that the convective diffusion rate of the ferrous ion is no longer the sole controlling factor. The ratio of measured-to-calculated current at 24 h for grey cast iron was found to be 0.508. In comparison, the ratio was 0.628 for Mondi A after 24 h at 1,000 rpm and 40°C (104°F), and 0.647 for ductile iron after 36 h at 1,000 rpm. Similarly, at 27°C (81°F) and over the velocity range used, grey cast iron showed lower currents than either Mondi A or ductile iron.

Upon applying the next rotation speed of 1,235 rpm, no change in current was observed. The corrosion current continued to decrease with time, although showing some instability. After 34 h, the measured-to-calculated currents ratio dropped to the very low value of 0.267. This is to be compared with 0.464 for Mondi A after 33 h at 1,235 rpm and 0.460 for ductile iron after 36 h. Hence, grey cast iron showed a greater attenuation of convective diffusion control than either alloy. Furthermore, this effect is more pronounced at higher temperatures, rotation speed, and exposure times. Similarly, the measured variation of corrosion potential with time and velocity for grey cast iron shown in Figure 4.10 indicated more noble values than that of either Mondi A and ductile iron. On initial contact, the corrosion potential assumed a value of -815 mV. It then increased to around -730 mV within 15 minutes, and continued to increase steadily to reach, after 49 h at 500 rpm, a value of -580 mV. By comparison, this value was, at 40°C (104°F), -710 mV after 59 h at 500 rpm for ductile iron and around -690 mV after 61 h at 500 rpm for Mondi A.

Upon increasing the rotation speed to 1,000 rpm, the corrosion potential showed a short transient but then resumed its continued increase to reach values of around -520 mV after 35 h. The corresponding values were -705 mV for ductile iron and -655 mV and -690 mV for the two Mondi A electrodes. At 1,235 rpm, it further increased slightly to stabilize at the more noble value of -500 mV.

This variation in corrosion potential, with time and velocity, is in line with the corresponding current behavior. A shift in potential in the noble direction is seen to be accompanied by a decrease in current. The decrease in current and the shift in potential were higher for grey cast iron, Mondi A, and ductile iron in that order.

4.4.5 Mondi B in Sulfuric Acid at 40°C (104°F)

This alloy, Mondi B, is the last in the present series of cast iron alloys tested in 98% sulfuric acid at 40°C (104°F) over a range of electrode rotation speeds. It was initially subjected to 500 rpm. Upon contact with the acid, the corrosion potential assumed a value of -800 mV and increased to -740 mV within 30 minutes, at which time it started to become unstable (cf. Figure 4.10). It did not stabilize again until after 12 h through the run and was then found to be around -690 mV. The measured current density was then 0.287 mA/cm^2. During this period of potential instability, the potentiodynamic scans could not be performed, and thus the data on the current were quite incomplete. A value of 0.400 mA/cm^2 at 5 h was obtained during a lapse in the potential instability. Figure 4.13 shows the variation of the measured and calculated currents with time and rotation speed, while Table D.33 in Appendix D lists the full set of results. In spite of these difficulties, it can safely be stated that the currents were initially very high and then started to fall very quickly. They subsequently decreased very slowly when they reached values comparable to or lower than that calculated.

Following the first test, the rotation speed was increased to 1,000 rpm, the current continued its steady decrease until after 59 h, the ratio of measured to calculated current was 0.498. This value is comparable to that found under similar conditions for grey cast iron (0.508) but lower than for either Mondi A (0.628) or ductile iron (0.647).

Finally, the rotating speed was further increased to 1,235 rpm, and after 51 h, the measured-to-calculated current ratio was 0.309. This value is comparable to that found for grey cast iron at 0.267 after 34 h at 1,235 rpm and 40°C (104°F). The corresponding values for Mondi A and ductile iron were 0.464 and 0.460, respectively.

Corrosion potential measurements (Figure 4.10) indicated a value of -664 mV after 51 h at 500 rpm. Upon increasing the rotation speed to 1,000 rpm, the potential showed similar values and stabilized at -666 mV. At 1,235 rpm, the potentials reached the more noble value of -638 mV after 51 h.

4.5 CONCLUDING REMARKS ON CORROSION OF CAST IRONS

The hydrodynamic corrosion of a number of cast irons has been investigated in 98% sulfuric acid at 27°C (81°F) and 40°C (104°F). A significant departure from the proposed convective diffusion model was observed, suggesting nonuniform corrosion and film formation as confirmed by visual and light microscopy examination. The graphitic residue, left behind upon corrosion of cast iron surfaces, is thought to form a second diffusional barrier that will lead to a decrease of the overall mass transfer process and of the corrosion current. These phenomena were observed to depend on the cast iron type, velocity, temperature, and exposure time. Furthermore, and particularly at 40°C (104°F), the observed decrease in corrosion current was accompanied by a reduction in the corrosion current sensitivity to flow and a shift of corrosion potential in the noble direction. Both effects were observed to vary with rotation speed, temperature, time, and the type of cast iron.

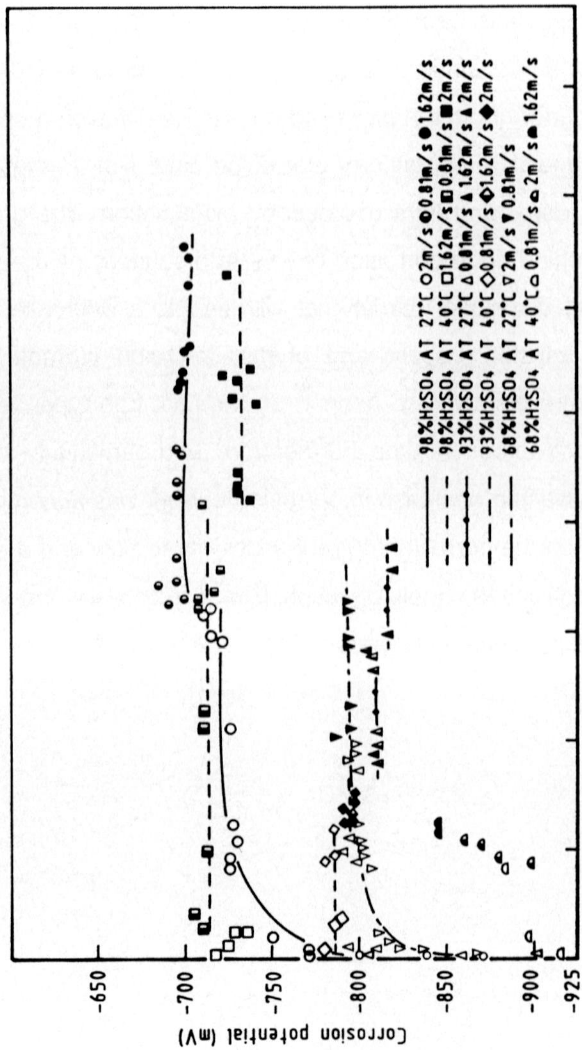

Figure 4.1 Variation of corrosion potentials with time and rotation velocity for mild steel under all conditions.

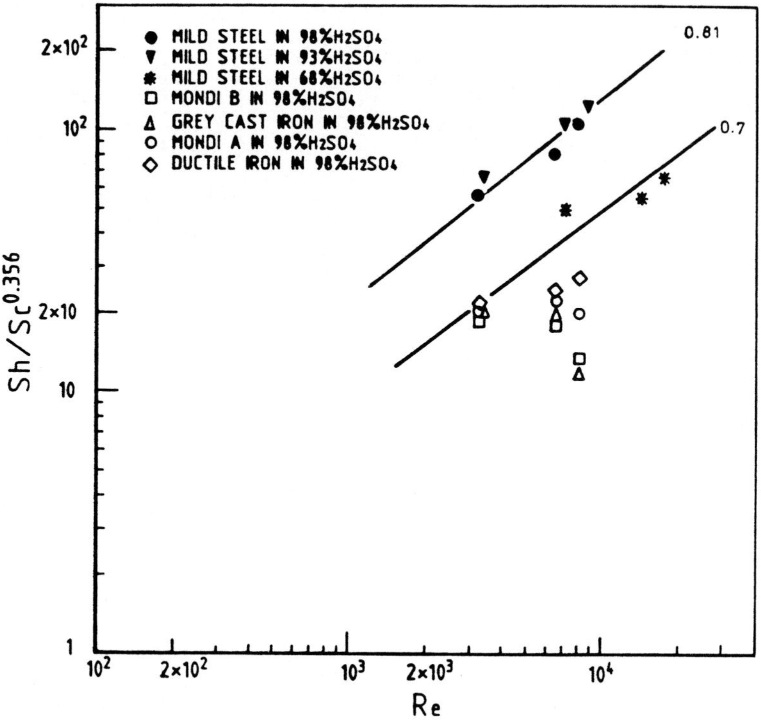

Figure 4.2 Plot of measured corrosion rates expressed as $Sh/Sc^{0.356}$ vs Re for all alloys tested at 40°C (104°F).

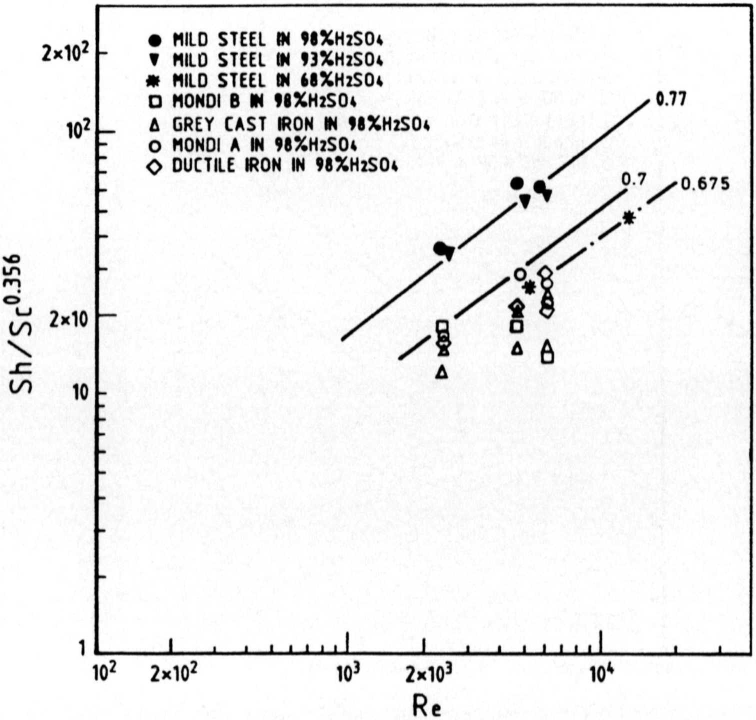

Figure 4.3 Plot of measured corrosion rates expressed as $Sh/Sc^{0.356}$ vs Re for all alloys tested at 27°C (81°F).

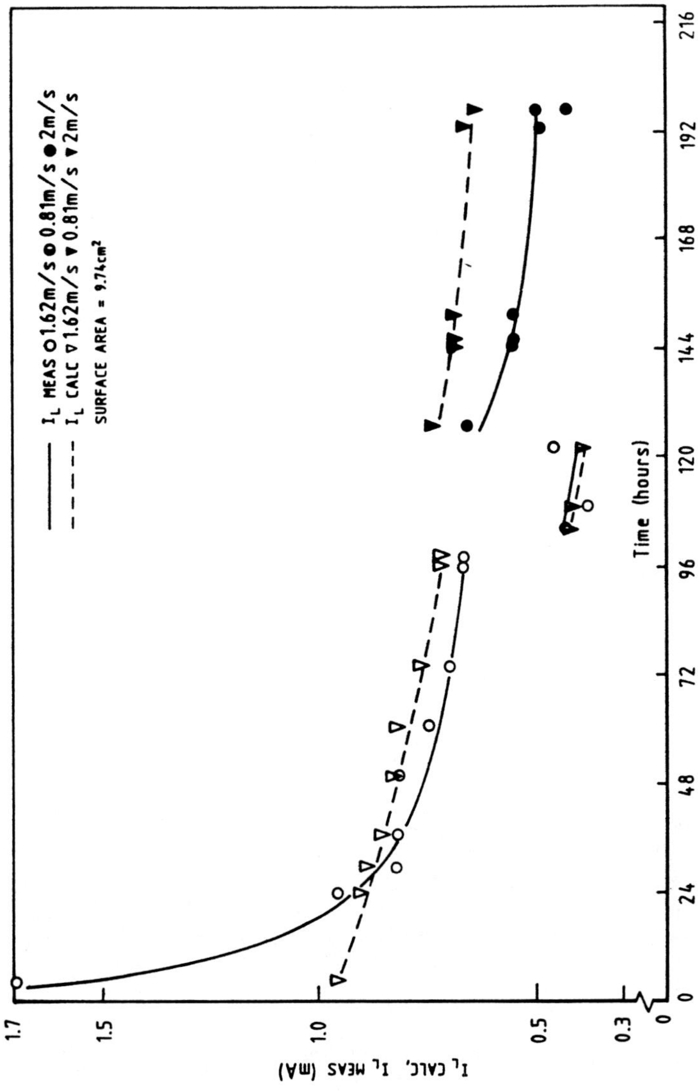

Figure 4.4 Variation of measured and calculated corrosion currents with time and rotation velocity for Mondi A in 98% H_2SO_4 at 27°C (81°F).

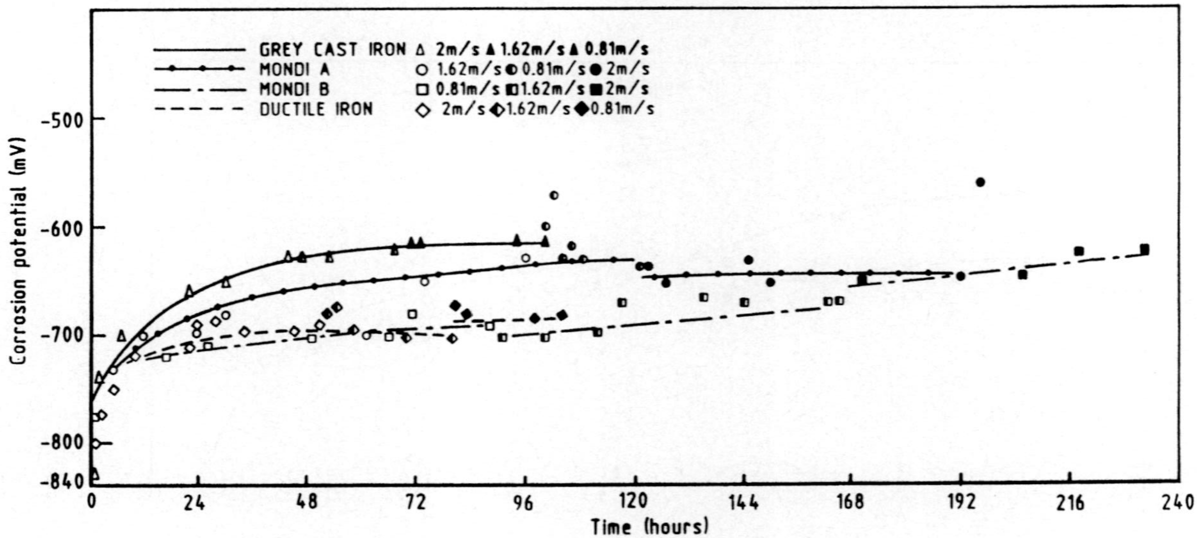

Figure 4.5 Variation of corrosion potential with time and rotation velocity for all four cast irons tested in 98% H_2SO_4 at 27°C (81°F).

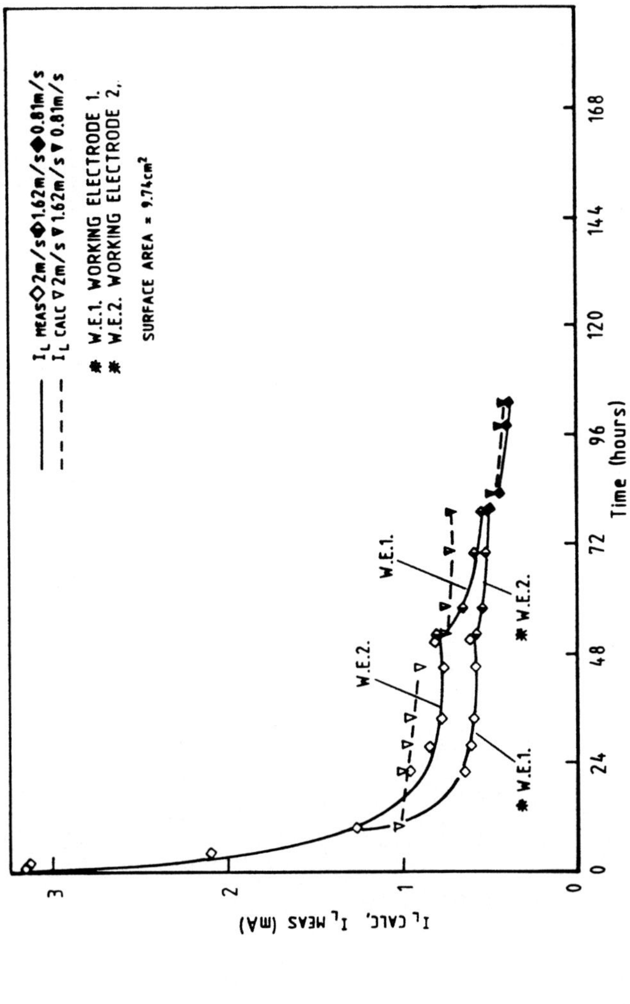

Figure 4.6 Variation of measured and calculated corrosion currents with time and rotation velocity for ductile iron in 98% H_2SO_4 at 27°C (81°F).

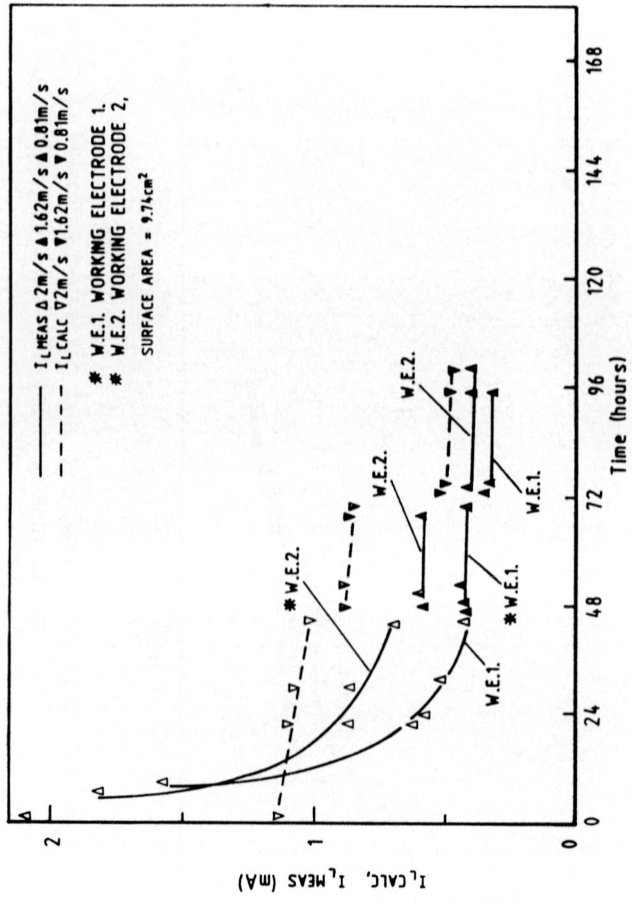

Figure 4.7 Variation of measured and calculated corrosion currents with time and rotation velocity for grey cast iron in 98% H_2SO_4 at 27°C (81°F).

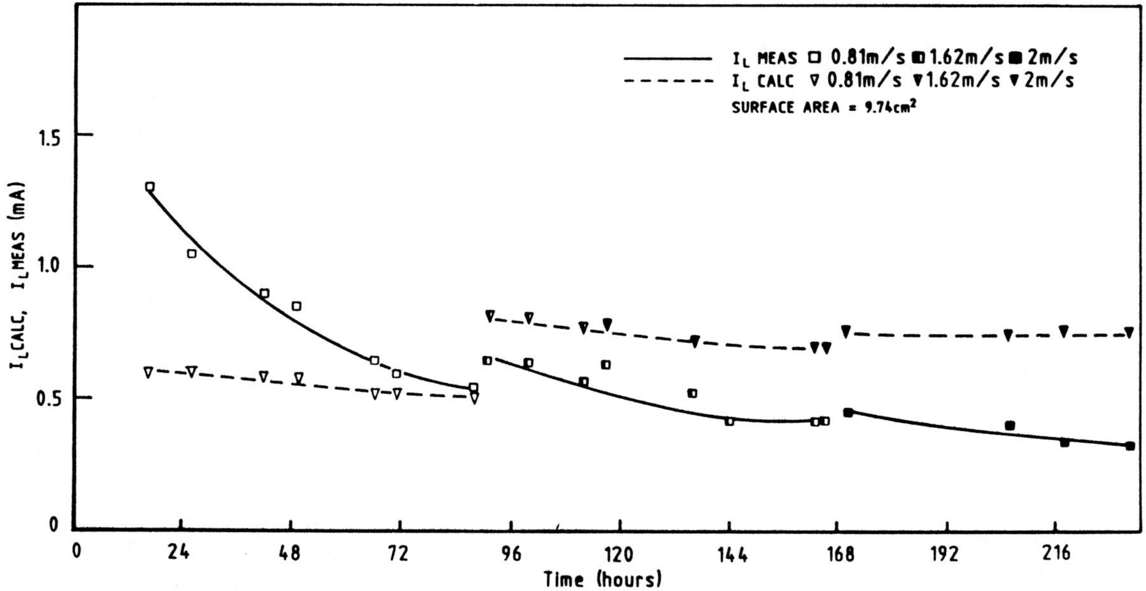

Figure 4.8 Variation of measured and calculated corrosion currents with time and rotation velocity for Mondi B in 98% H_2SO_4 at 27°C (81°F).

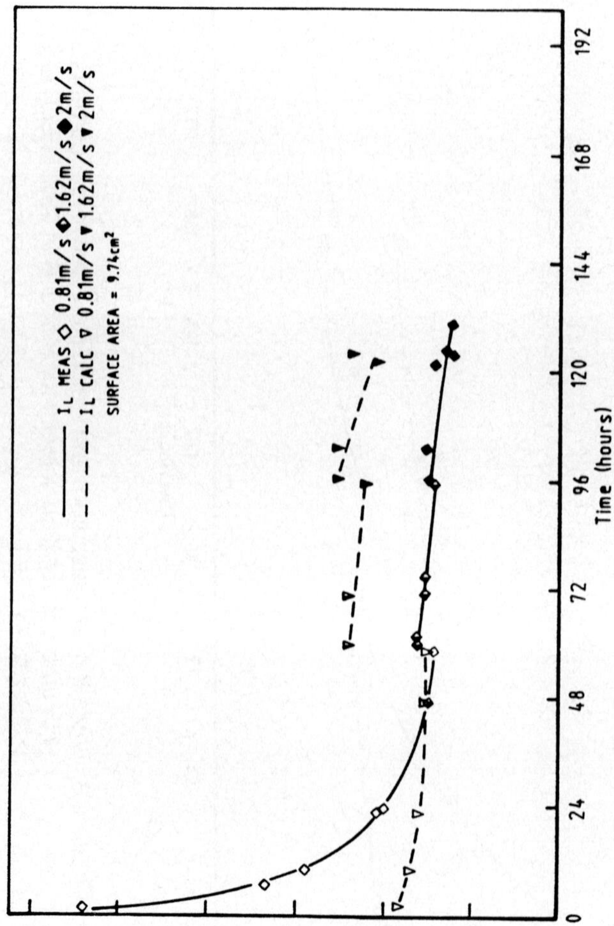

Figure 4.9 Variation of measured and calculated corrosion currents with time and rotation velocity for ductile iron in 98% H_2SO_4 at 40°C (104°F).

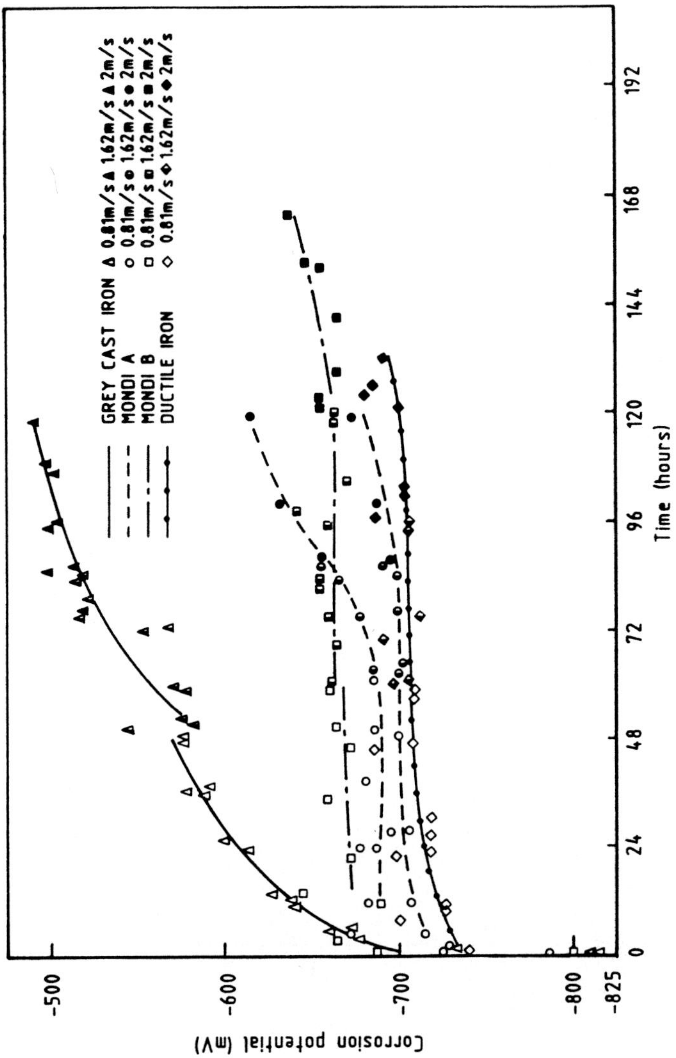

Figure 4.10 Variation of corrosion potential with time and rotation velocity for all four cast irons tested in 98% H_2SO_4 at 40°C (104°F).

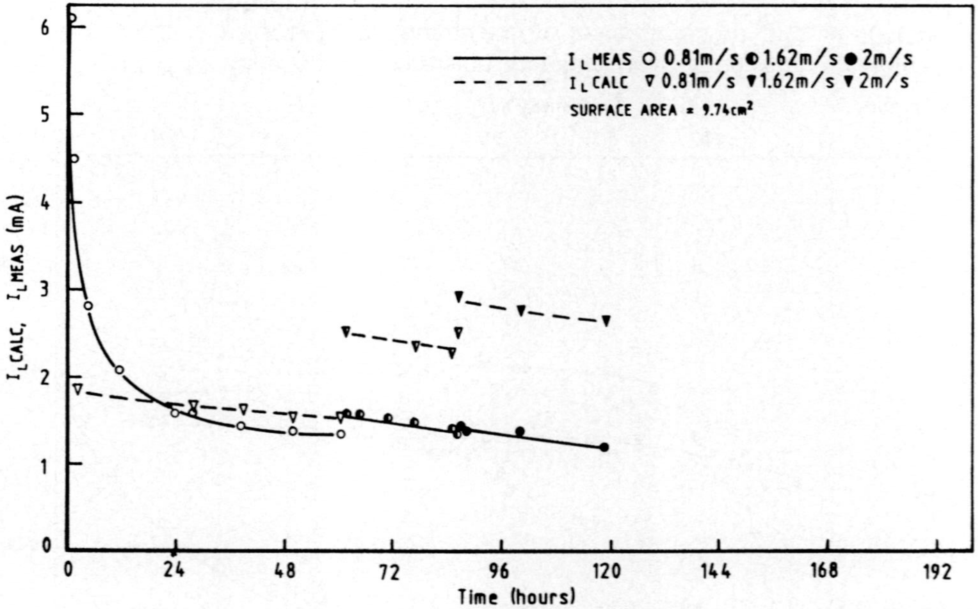

Figure 4.11 Variation of measured and calculated corrosion currents with time and rotation velocity for Mondi A in 98% H_2SO_4 at 40°C (104°F).

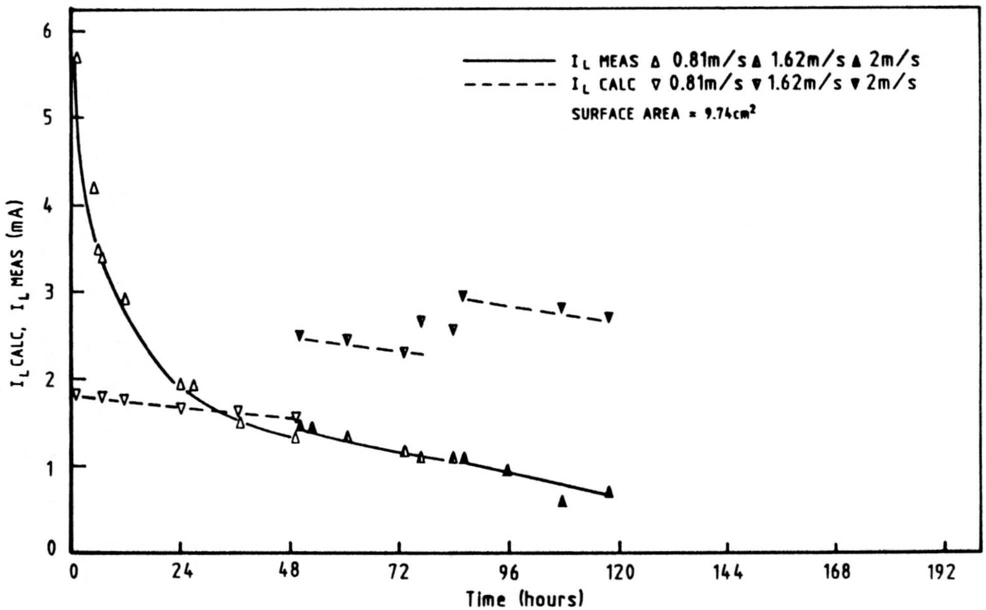

Figure 4.12 Variation of measured and calculated corrosion currents with time and rotation velocity for grey cast iron in 98% H_2SO_4 at 40°C (104°F).

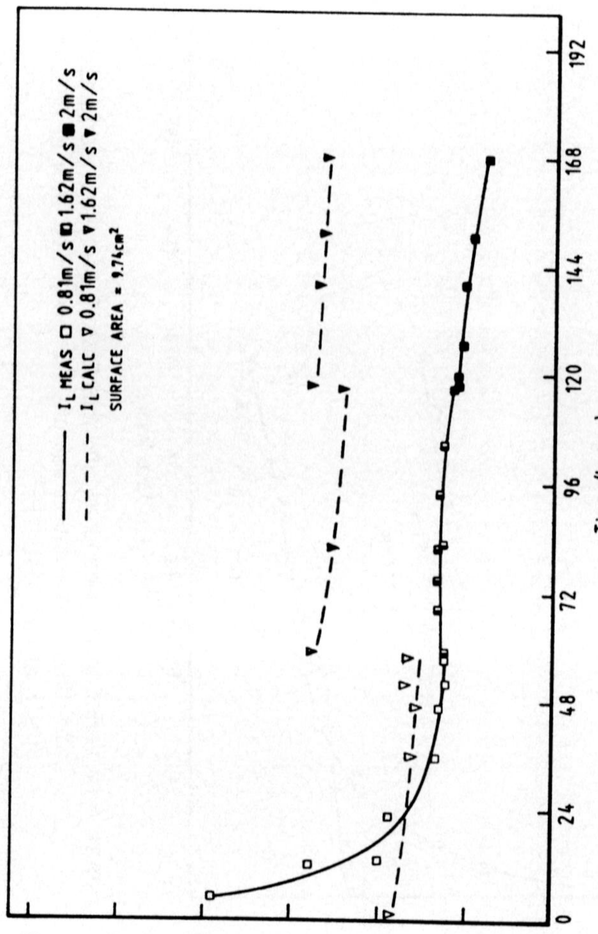

Figure 4.13 Variation of measured and calculated corrosion currents with time and rotation velocity for Mondi B in 98% H_2SO_4 at 40°C (104°F).

CHAPTER
FIVE

ANALYSIS AND DISCUSSION

5.1 CORROSION OF MILD STEEL

5.1.1 Introduction

The corrosion of mild steel in 68%, 93%, and 98% sulfuric acid at 27°C (81°F) and 40°C (104°F) and over the 0.81 to 2 ms^{-1} velocity range has been found to be controlled by the convective diffusion of ferrous ion away from a ferrous sulfate saturated layer-liquid interface. Whereas the general features observed for all conditions tested were similar, i.e., the corrosion increases with rotation velocity and decreases with bulk acid iron contamination, the mass transfer rate was not proportional to the velocity (Reynolds numbers [Re]) raised to a constant power of 0.7. The mass transfer dependence on the 0.7 power of velocity has been indicated in the literature to hold for the 0 to 99% acid concentration range.[9,76] This study arrived at a value for the velocity exponents of 0.77 in 98% and 93% sulfuric acid at 27°C (81°F) and 0.81 at 40°C (104°F) (Figures 4.1 and 4.2). These new exponent values have been found by assuming that the constant 0.079 and the Schmidt (Sc) exponent of 0.356 in the Eisenberg relation are true.

5.1.2 The Origins of the Eisenberg Relation

Eisenberg et al.[46] used the Chilton-Colburn analogy:[22]

$$j_D = \frac{K}{V} Sc^{\frac{2}{3}} = f(Re) \tag{5.1}$$

in their empirical derivation of a mass transfer relation applicable to turbulent flow at a rotating cylinder electrode (RCE). The exponent ⅔, employed by Chilton and Colburn, was not based on diffusion data;[22] it was chosen by a successful correlation of early heat transfer data. Mass transfer data were scarce at that time and represented systems with Sc values close to unity. A substantial extension of the Sc range to values of 1,000 to 3,000 was made by Linton and Sherwood who studied dissolution of solids in flow through round tubes, past flat plates, and other shapes.[21] The Eisenberg study includes Sc numbers from 835 to 11,490. The simplicity of the correlating parameter, j'_D, found:

$$j'_D = \frac{K}{V} Sc^{0.644} = 0.079 Re^{-0.3} \tag{5.2}$$

and its closeness to the Chilton-Colburn parameter, j_D, is rather remarkable in view of the uncertain empirical basis of the latter.

The exponent, ⅔, on the Sc number is frequently subjected to theoretical rationalization, especially using boundary layer theory.[99] Chilton is sometimes said to have cheerfully conceded that the value of ⅔ was not even equal to the best fit data but was chosen because the slide rule in those days (1934) had square root and cube root scales but no other easy way to take exponents.[99]

5.1.3 Discussion of the Present Work

The Ellison and Schmeal results in 68% sulfuric acid that have been reproduced in this study were obtained at two values of the Sc number: 71,686 and 36,018 at 27°C (81°F) and 40°C (104°F), respectively.[9] In 93% and 98% sulfuric acid, the Sc values are almost one order of magnitude higher. Table 5.1 gives the different values of the kinematic viscosity, diffusion coefficient, and Sc number encountered at the two temperatures used.

Before discussing any possible relation between the higher Re exponents found in 93% and 98% sulfuric acid and the Sc range, it is worth remembering that the Sc number is not a function of flow velocity (or electrode rotation rate) or of the characteristic dimensions. It is determined exclusively by the physical properties that characterize momentum transfer and the transport of matter by a purely molecular mechanism. As the kinematic viscosity increases, the diffusion coefficient decreases in accordance with the relation $D = CONST/\nu$ (cf. Sections 1.2 and 3.3). Thus, the Sc number increases as the square of the kinematic viscosity. In viscous fluids, the Sc number attains a value of 10^6 and higher. A high value of Sc number signifies that, even at very low velocities, the transport of matter in a fluid by convection predominates over molecular diffusion. Also, the smaller the diffusion coefficient, the closer to the wall is the region in which turbulent eddies predominate over molecular diffusion in mass transfer.

Examination of Table 5.1 shows the Sc values (174,456 to 412,440) for 93% and 98% sulfuric acid to be well outside the range (835 to

11,490) used by Eisenberg in establishing a mass transfer relation for turbulent flow at an RCE. They were similarly above the Sc values for 68% sulfuric acid concentration at which the relation was also found to apply. However, the higher Re exponents at higher temperatures, i.e., lower Sc numbers, in both 93% and 98% sulfuric acid make arriving at any conclusion regarding the effect of the Sc number range on the Re exponent quite impossible. This may also mean that the applicability of the relation cannot be extended to Sc values much higher than those previously reported.

Any possible influence of the physical characteristics (Sc number) of the dissolution mechanism on the hydrodynamics (Re number exponent) under which mass transfer occurs can be investigated further if a range of acid concentrations from 50% to 98% at different temperatures is used, thus giving wide ranging Sc numbers. However, this is well beyond the scope of this work, which was to use the RCE to generate hydrodynamic corrosion data for mild steel (and cast irons) in 93% and 98% sulfuric acid.

The experiment suggested above can also be planned to give a better definition of the Sc exponent; this can be achieved by plotting K/v against Sc to find a set of lines at constant Re. The slope of such lines then gives the Sc exponent.

Another possible explanation for the observed increase in mass transfer rate in 93% and 98% sulfuric acid, as compared to 68%, resides in the nature of the ferrous sulfate film. This has been found to be $FeSO_4 \cdot H_2O$ in 68% sulfuric acid and $FeSO_4$ in 93% and 98% sulfuric acid.[3,95] There is some evidence to suggest that $FeSO_4$ is a less adherent form of the ferrous sulfate film.[90] S. Dean reports that above approximately 99% acid concentration, corrosion rates of mild steel are higher than those predicted from a ferrous sulfate diffusion model, and therefore, other mechanisms are also operating in this range.[90,100] There may be some crystallographic or structural changes that occur in ferrous sulfate that make it lose its adhesion to the steel. If such a transformation is possible, our results could be considered to suggest that this happens at a lower concentration than 99%.

The extrapolation formula suggested by Ellison and Schmeal used throughout this research to calculate diffusion coefficients, and hence, predict corrosion currents (cf. Section 3.3), can also be a source of discrepancy.[9] The formula was arrived at through measurements of the diffusion coefficient for ferrous ion in 68% sulfuric acid using a rotating platinum disc electrode. Although its use in 68% sulfuric acid led to agreement between the model and the experimental results, it is possible that the formula under-predicts the diffusion coefficient of ferrous ion in acids of higher concentration. The unexpected results of this study would suggest that any future work should endeavor to measure the diffusion coefficient of ferrous ion in the actual condition tested.

5.1.4 Comparative Discussion of Previous Work

A number of workers have found results showing a significant departure from the Eisenberg correlation. Hydrodynamic considerations enabled Frank-Kamenistii to show that, for the limiting case of large Re and Sc numbers, the rate of mass transfer is linearly related to the velocity.[101] Earlier work by King et al. showed that at low rotation speeds, n = 0.7. However, as the rotation rate increased, a linear variation was approached, and, in some cases, no correlation was apparent.[65,66]

These wide-ranging variations in the Re exponent values are due to complicating effects, such as contribution of bubble formation and evolution,[58] partial diffusion control,[28-30] and surface roughening during anodic dissolution (cf. Section 2.2.4). The Eisenberg relation, which applies to pure convective diffusion at a smooth cylinder, is not expected to apply to those situations.[86] However, a critical review of the literature reveals more modest variations in the values of the power exponents; for example, the exponent for the Re number has been reported as 0.7,[31,46,63] 0.69,[62] 0.667,[102] and 0.60,[103] while the Sc exponent values reported include 0.41,[62] 0.356,[31,46,63,104] and 0.334.[103] Makrides made a valuable contribution in his analysis of a number of workers' results.[60] In the following paragraphs, a comparison of his findings and of the present results is made.

The Re exponent depends on the Re number. Makrides found that the velocity exponent value of 0.7, reported by Eisenberg et al., is a best value for the 1,000 to 100,000 range of Re numbers.[46] The exact value was 0.68 for the interval Re = 1,000 to 10,000, while it was 0.77 for Re = 30,000 to 100,000. The present study covered a more limited range of Re values: 2,360 to 17,873, and it is unlikely that these values would have any effect on the velocity exponent. This is particularly true because in 68% sulfuric acid, Re values ranged from 5,282 to 17,873, and the Re exponent was found, within experimental error, to be 0.7.

A least squares analysis of the results published by Eisenberg has been conducted to find that the limiting rates for reduction and oxidation depend on the 0.72 and 0.77 power of velocity, respectively.[60] The higher rates found for oxidation were attributed to surface roughening caused by the dissolution process. Following dissolution, the mild steel specimens used in the work described in this report did not show any surface roughening that could have led to the observed increase in the velocity dependence.

Differences in geometric arrangements, such as the length-to-diameter ratio, do not appear to be significant. Eisenberg et al., found no effect for an almost four-fold change in this ratio.[46] Baffling also has no effect as Eisenberg found no dependence of the gap between the rotating cylinder and the outer concentric electrode on mass transfer. These observations support the view that when mass transfer is controlled by the hydrodynamic conditions near a wall, the presence of other surfaces at a distance of the order of 1,000 to 10,000 times the thickness of the diffusion boundary layer, in a concentric or eccentric arrangement and the extent of end-baffling (cf. Section 2.2.2), are not expected to influence the rate of mass transfer.[60] The agreement between the results of this study in 68% sulfuric acid and that of Ellison and Schmeal adds to the evidence that geometrical variations have no influence on the mass transfer mechanism. Furthermore, these conclusions substantiate the decision to use an auxiliary electrode mounted on the rotating shaft as opposed to an outer concentric electrode. As far as the author is aware, the latter has been exclusively used in work reported in the open literature. In view of the design difficulties, the corrosivity of concentrated sulfuric acid and the high cost of a platinum-mesh cylindrical counter electrode large enough to fit the cell used (cf. Section 3.1), this last point together with the geometrical considerations listed above can prove to be of utmost practical significance for the design of an RCE apparatus to be used in industrial applications.

Finally, the cell used in this study could not be successfully sealed, and some vortex shedding was thus present at the highest velocity. In view of the agreement between our results and those of Ellison and Schmeal in 68% sulfuric acid, it is tempting to assume that the bubble formation observed had no effect on the mass transfer process. Furthermore, the source of the sealing problem was in the friction between the running seal and the stainless steel rotating shaft, leading eventually to the opening up of the seal. The exact reason for this happening is not understood, but the problem could be cured if a rotating shaft in polytetrafluoroethylene (PTFE) is fitted. This is particularly true because the running seal is in graphite-filled PTFE, and no major friction should be expected at their interface.

5.1.5 Discussion of Previous Semitheoretical Derivations of a Mass Transfer Correlation Applicable to the RCE

In their semitheoretical treatment of mass transfer under turbulent flow conditions at an RCE, a summary of which has been given in Appendix A, Gabe and Robinson arrived at the following mass transfer correlation:[13]

$$Sh = (2b)^{\frac{1}{3}} Re^{\frac{2}{3}} Sc^{\frac{1}{3}} \tag{5.3}$$

The constant ($[2b]^{1/3}$) must be determined experimentally. The argument used clearly involves a number of approximations, and the exponents found for Re and Sc can be affected by modified assumptions. These modifications would primarily concern the expressions postulated for the eddy viscosity expressions in the developed turbulence zone and in the viscous sublayer zone. The authors themselves, aware of the limitations of their derivation, point out the effect of modified hypotheses on the exponents of the Re and Sc numbers in Equation (5.3). Nevertheless, they conclude by saying that the analysis yields an equation of the correct form, and any subsequent treatment must be based on a refined model. The reader is referred to the original paper by Gabe and Robinson for more details.[13]

The Von Karman three-zone model used by Gabe and Robinson has been questioned by Levich, who postulated the addition of a fourth layer; namely, the turbulent boundary layer.[105,106] In this zone, neither molecular viscosity nor molecular diffusion play a noticeable part. This addition has been advocated to give a smoother transition from one layer to another than the Von Karman model. The Levich treatment arrives at a different expression for the coefficient of turbulent diffusion in the viscous sublayer:

$$\epsilon = D_{TURB} \simeq \nu\left(\frac{y}{\delta h}\right)^4 \tag{5.4}$$

This in turn would yield a value of ¼ for the Sc exponent in the mass transfer correlation at an RCE. To date, the argument has not been resolved because no single semitheoretical or theoretical approach has yielded a completely satisfactory solution, and empirical derivation of mass transfer relations for all but the very simple flow geometries is still heavily relied upon.[37-39,44,45]

5.1.6 The Theory of Laboratory Simulation in Hydrodynamic Corrosion Research and Its Application to the Mild Steel/Sulfuric Acid System Using the RCE

5.1.6.1 Introduction to the General Situation

The primary goal of this research has been to generate hydrodynamic corrosion data on mild steel and cast irons in concentrated sulfuric acid using the RCE geometry. However, there is always the need to recalculate results obtained in the laboratory to geometries that are both practical and more directly related to corrosion in a plant environment. The most important application, the pipe flow, has been rather extensively used,[9,12,75] and its mass transfer characteristics are now well understood.[68,72-74]

Although pipe flow geometry is directly related to practical applications, it tends to use cumbersome and bulky apparatus with leak containment provisions and extensive facilities for temperature and flow control. Rotating electrodes give a more compact and easily handled apparatus and are therefore to be preferred in the initial generation of corrosion data. The use of the RCE is also advantageous in that it operates under well-defined hydrodynamics and uniform current distribution (cf. Section 2.2.2). Furthermore, it operates under turbulent conditions at all but the very low velocities.[19,46] This is highly significant since pipe flow in plant condition is usually turbulent and also because the geometries to be compared must operate under the same flow regime, i.e., both turbulent or both laminar. It would be erroneous, for example, to use a rotating disc operating under laminar flow conditions to simulate a pipe operating under turbulent conditions. If the mechanism responsible for corrosion is mass transfer, then it is possible to equate the mass transfer relation of the RCE to that, if known, of the geometry to be simulated. If the RCE is operated to obtain results that are to be recalculated for pipe flow, then it is possible to write:

$$Sh_{RCE} = Sh_{PIPE} \tag{5.5}$$

This procedure has been used by a number of workers.[20,75] It assumes that the effect of flow in the laboratory geometry (i.e., RCE) models the effect of flow in pipes when the mass transfer rates in the two geometries are equal. The above relation is solved so that the pipe velocity is expressed in terms of the cylinder rotation rate. The method has been successfully used to predict corrosion rates of Cu-Ni alloys in oxygenated 1 M NaCl solutions by Chen et al. using the RCE, flow pipe, and the annular flow channel geometries.[75]

5.1.6.2 The Shear Stress Criterion Hypothesis

More recently, Silverman has suggested that the method of quantitatively relating the mass transfer relations must also ensure that the interaction between the alloy surface and the transfer of momentum is equivalent in the two geometries.[107] In his opinion, laboratory simulation then has a reasonable chance of duplicating the velocity-sensitive mechanism found in the plant geometry for the same alloy and environment. The shear stress is one such measure of the alloy surface-fluid interaction; hence, the apparatus used to simulate the plant geometry should be operated so that the wall shear stress range is the same as that encountered in the plant geometry. The shear stress at the wall can be estimated for many geometries so that the following equation may be written:[31,108]

$$\tau_{LABORATORY} = \tau_{PLANT} \tag{5.6}$$

If this equation is fulfilled, then for a given system, the mechanism by which fluid velocity affects corrosion rate in the plant is proposed to be identical to that which affects corrosion rate in the laboratory. Only then can the appropriate corrosion rate-fluid mechanics relationship (i.e., mass transfer correlation) for the plant geometry and for the same mechanism be used to determine the expected corrosion rate in the plant.

Support for these arguments can be found in Silverman's own experimental work,[107] as well as in the work of Ellison and Schmeal,[9] who showed that a mass transfer controlled model for carbon steel corrosion in sulfuric acid explains the data for both pipe and rotating cylinder geometries.

Fluid velocity can also affect the corrosion rate through erosion. Efird has shown in pipe flow of seawater through copper-nickel alloys that once the shear stress becomes greater than a critical value, erosion accelerates the corrosion rate.[18] Shear stress seems here to be the governing variable. This point led Efird to conclude that the concept of critical shear stress could be applicable to other geometries handling seawater. If both geometries operate in the same turbulent or laminar flow regimes, the above argument suggests that the breakaway shear stresses in both geometries could be comparable for the same alloy. Indeed, if erosion in other alloy-environment systems depends only on shear stress, then fulfilling the shear stress conditions while maintaining the same flow regime may allow one geometry to model erosion-corrosion mechanistically for another geometry.

In addition to the experimental support found for the Silverman hypothesis,[9,18,107] there is evidence as to its validity from the Chilton-Colburn analogy between heat, momentum, and mass transfer:[22]

$$\frac{Sh}{Re \cdot Sc} = \frac{f}{2} g(Sc) \qquad (5.7)$$

In this equation, the empirical friction factor (f) that quantifies the momentum transfer (velocity profile), which gives rise to shear stress, may be related directly to the Sherwood number (Sh) that contains the mass transfer coefficient.

This similarity suggests that the equating of shear stresses may be a reasonable method of enabling the momentum transfer mechanisms to have a similarity in the two geometries.

In conclusion, and by the analogy of Equation (5.7), the similarity in momentum transfer suggests a comparable similarity in the mass transfer mechanism for a given fluid-alloy system in different geometries.

5.1.6.3 The Mild Steel Concentrated Sulfuric Acid System

An earlier investigation into this system at the 68% acid concentration level was made by Ellison and Schmeal using the RCE and the pipe flow geometries.[9] Silverman (previous section) pointed out that in this case, the range of shear stresses used in the pipe geometry overlapped the range of shear stresses used in the RCE geometry (however, this point was not raised in the original paper by Ellison and Schmeal). Since complete mass transfer control was found in both geometries, the equating of the mass transfer relationships is then possible:

$$Sh_{RCE} = Sh_{PIPE} \tag{5.8}$$

$$0.079Re^{0.7}Sc^{0.356} = 0.0096Re^{0.913}Sc^{0.346} \tag{5.9}$$

Equation (5.9) is then solved so that the pipe velocity is expressed in terms of the cylinder rotation rate. Therefore, the pipe flow would be simulated by a rotating cylinder.

In this study, the results of Ellison and Schmeal in 68% sulfuric acid have been confirmed, and further work in 93 and 98% sulfuric acid showed a range of velocity exponent values of 0.77 to 0.81. This means that at those concentrations, the left-hand side of Equation (5.9) needs to be modified accordingly.

Ellison and Schmeal have found the relation for turbulent mass transfer in pipe flow derived by Harriot and Hamilton[68] to be valid for mild steel in 68% sulfuric acid. However, there are very few data at higher acid concentrations for the pipe flow geometry, and as the RCE results showed some dependence of the velocity exponent on acid concentration, the possibility that this may also be the case in the pipe flow geometry cannot be ignored.[10,12] Hence, the need to study corrosion of mild steel in acids of the 93% and 88% concentration range in this geometry arises. The likely result of such an investigation is a probable modification of the right-hand side of Equation (5.9) as a result of the dependence of the Re number exponent on acid concentration in the mass transfer correlations. (This is only a hypothesis for the pipe flow geometry.) It is also likely, because of the momentum-mass transfer analogy, that the expression for the friction factor (i.e., the resistance to fluid flow created by the wall) in the individual geometries, and thus, those for the shear stress, will be modified.[22,31,108] These points may also need further investigation if new work in 93% and 98% sulfuric acid in the pipe geometry is envisaged.

5.1.6.4 Concluding Remarks About the Transfer of Corrosion Rate to More Complex Geometries

The use of pipe flow geometry, or of the RCE results recalculated for the pipe flow, still leaves the need for a method of interfacing the pipe loop results with other geometries such as pumps, reactors, and tanks. However, most workers have reported work on cylinders, discs, and pipes and suggested empirical correlations or correlations based on an equality of mass transfer for the transfer of corrosion rate.[10,20,109] For this part, Poulson has extended the technique by using a flow loop in which specimens of different geometries have been fitted. These are, namely, the rotating cylinder within a cylinder, the rotating disc, the impinging jet, a nozzle or orifice, and a straight pipe section.[15] (However, no indication is given of either the metal-environment used or of the results he may have found.) In this way, the influence of geometry on the corrosion characteristics of a particular alloy-environment in which flow sensitivity has been found (i.e., cases of complete or even partial mass transfer control of corrosion) can be studied.[107] This type of approach will be difficult, and hence, will become even more empirical when extended to geometries such as pump impellers, multiple plates in pipe for which mass transfer characteristics are not known.

An example of the difficulties introduced by complex geometries in sulfuric acid environment has been commented on by S. Dean et al.[12] The corrosion rates measured in elbows has been found to be 3 to 4 times the straight pipe rate as predicted by the mass transfer correlation (right-hand side of Equation [5.9]). This has often been referred to as erosion-corrosion when, in reality, it is due to enhanced mass transfer. Localized heat transfer rates in turbulent flow have been measured in elbows and found to be as much as 3.5 times the rate in straight pipes for turbulent flow.[110] By the Chilton-Colburn analogy, the same enhancement factor for mass transfer, and therefore corrosion, should be expected and has indeed been found.[12,22] To minimize this effect, long radius elbows are often specified for sulfuric acid pipelines. Similarly, higher corrosion rates than predicted by the convective diffusion model will be found for tees, valves, and internal surface roughness caused, for example, by improper pipe alignment during welding.[12]

5.2 CORROSION OF CAST IRONS

5.2.1 Introduction

The corrosion of the range of cast irons studied in 98% sulfuric acid at both temperatures used (namely, 27°C [81°F] and 40°C [104°F]) and over the 0.81 to 2 ms^{-1} velocity range showed a significant departure from the proposed convective diffusion model. This was in the form of a reduction in corrosion current below the values predicted by the model and a partial-to-total loss of the corrosion current flow sensitivity. Both effects were observed to depend on temperature, rotation velocity, cast iron type, and exposure time.

Cast irons have been reported to corrode with the appearance of a graphitic residue that leads to stifling of further corrosion;[98] scanning electron microscopy (SEM) of the corroded surfaces has confirmed the existence of surface films on the exposed cast iron specimens.

This evidence and the hydrodynamic corrosion data have led to a complete reappraisal of the proposed model to introduce ways to account for the variety of current reduction levels and associated phenomena observed.

5.2.2 The Filming Factor—P

The diffusion-controlled limiting current predicted by the model is given by:

$$I_{L\ Calc} = zFK_{Calc}(c_s - c_b)S \tag{5.10}$$

since the mass transfer factor $K_{CALC} = \frac{D}{\delta d}$, the expression for $I_{L\ CALC}$ can be rewritten as:

$$I_{L\ Calc} = zF\frac{D}{\delta d}(c_s - c_b)S \tag{5.11}$$

where δd, the diffusion boundary layer thickness, is the region in the model where most of the resistance to mass transfer occurs.[13] However, the results of this study on cast irons were quite unexpected. As a result, an empirical approach has been adopted. A new entity, named the filming factor (P), was introduced to account for the lower currents observed. P was defined as follows:

$$I_{L\ MEAS} = \frac{zF(c_s - c_b)S}{\frac{\delta d}{D} + P} \tag{5.12}$$

Equation (5.12) shows P to have the inverse dimension of a mass transfer factor. Figure 5.1 is a plot of the calculated values of P with time for all four cast irons in 98% sulfuric acid at 40°C (104°F), while Figures

5.2 and 5.3 show the variation of the calculated P values with velocity for all cast irons studied in 98% sulfuric acid at 40°C (104°F) and 27°C (81°F), respectively. Examination of Figure 5.1 shows that at longer exposure times and thus lower corrosion currents (cf. Appendix D, Tables D.30 through D.33), the increase in P shows features distinctive to the alloy type. For Mondi B and grey cast iron, which exhibited very low currents, the values of P were very high and did not show any sign of stabilization. As for the other two alloys, Mondi A and ductile iron, the P values tended to level off, suggesting that some form of steady state of the corrosion mechanism has been reached. Since the corrosion currents were below the calculated values by a higher proportion at 40°C (104°F) than at 27°C (81°F), the P values were also higher at the higher temperature (cf. Figures 5.2 and 5.3).

It can clearly be seen from Equation (5.12) for the measured current that a higher value for P will be associated with an overall higher resistance to mass transfer than predicted by Equation (5.10) for the calculated limiting current. Also, the combination of Equations (5.10) and (5.12) gives the following expression for P:

$$P = \frac{1 - X}{K_{CALC} \cdot X} \qquad (5.13)$$

$X = \frac{I_{L\,MEAS}}{I_{L\,CALC}}$ designates the ratio of the measured-to-calculated currents. The values of P calculated for all cast irons in 98% sulfuric acid at 40°C (104°F) are reported in Appendix D, Tables D.30 through D.33, while Table D.35 of the same Appendix gives the P values calculated for the cast irons in 98% sulfuric acid at 27°C (81°F).

The approach suggested in this section, though essentially empirical, has identified P, the filming factor, with an increase in mass transfer resistance. This increase must be associated with a second physical diffusional barrier. This can only be the presence of a film of retained corrosion products on the surface.

Substantiation of the existence of a surface film can also be found in the increasing values with time of the corrosion potentials for cast irons at 27°C (81°F) and 40°C (104°F) as shown in Figures 4.5 and 4.10, respectively. For example, grey cast iron at 40°C (104°F) with its low corrosion currents and complete loss of flow sensitivity showed corrosion potentials as high as -500 mV. This would suggest severe filming conditions (cf. also section 1.5).[32] Between -600 mV and -700 mV, the corrosion potentials at 27°C (81°F) were generally less high and the corrosion currents depressed by a lesser proportion below the calculated values.

5.2.3 The Mass Transfer Characterization of the Surface Film

The filming factor (P), which has the inverse dimension of a mass transfer factor, represents the added mass transfer resistance due to the presence of a corrosion products film. Bearing this in mind, Equation (5.12) can be written as:

$$I_{L\,MEAS} = \frac{zF(c_s - c_b)S}{\frac{\delta d}{D} + \frac{e}{Df}} \tag{5.14}$$

where the term $\frac{e}{Df}$, which replaces P, is the inverse of the mass transfer factor through the surface film; e is the thickness of the corrosion products film; and Df is the diffusion coefficient of ferrous ion, the reacting species, through the corrosion products.

By replacing $\frac{D}{\delta d}$ by K_{CALC} and defining the mass transfer factor (K_{FILM}), through the film as $\frac{1}{K_{FILM}} = \frac{e}{Df}$, Equation (5.14) will then become:

$$I_{L\,MEAS} = \frac{zF(c_s - c_b)S}{\frac{1}{K_{CALC}} + \frac{1}{K_{FILM}}} \tag{5.15}$$

If the overall resistance to mass transfer is to be expressed by a newly defined factor, the measured mass transfer factor, K_{MEAS}, Equation (5.15) will become:

$$I_{L\,MEAS} = \frac{zF(c_s - c_b)S}{\frac{1}{K_{MEAS}}} \tag{5.16}$$

By combination of Equations (5.15) and (5.16), it is found that:

$$\frac{1}{K_{MEAS}} = \frac{1}{K_{CALC}} + \frac{1}{K_{FILM}} \tag{5.17}$$

Equation (5.17) simply states that the measured overall resistance to mass transfer in cast irons is the sum of the resistance through the diffusion boundary layer and of the resistance through the retained corrosion products on the surface.

Since the corrosion behavior observed experimentally showed a strong dependence on cast iron type, temperature, and time, a study of the variations of K_{MEAS} and K_{FILM} with the parameters of interest: cast iron type, temperature, velocity, and exposure time, together with the SEM observations of the corroded surfaces, should be revealing of the corrosion mechanism of the cast irons studied. This is the subject of the next section.

5.2.4 Discussion of the Observed Corrosion

In Chapter Four, Sections 4.3.1 and 4.4.1, it has been reported that visual examination of the exposed cast iron specimens showed the surfaces to have become roughened as a consequence of corrosion. The extent of this roughness was higher at 40°C (104°F) than at 27°C (81°F). It was also greater for grey cast iron, decreasing through Mondi B, ductile iron, and Mondi A, in that order. Therefore, it was decided to examine the corroded surfaces by SEM to reveal the microscopic features of this phenomenon of surface roughening. SEM should also confirm the existence of, and reveal the nature of, the surface films that are supposedly responsible (cf. Section 5.2.3) for the varying degree of corrosion-current reduction and for the loss of flow sensitivity.

The visual observations of the surface roughening were more or less substantiated by the series of micrographs produced by the SEM. For example, grey cast iron shows a very uneven surface covered to a large extent by protruding graphite flakes. Both effects were more pronounced at 40°C (104°F) than at 27°C (81°F) (Figures 5.6 and 5.7, respectively). At the other end of the roughness range, Mondi A and ductile iron at 27°C (81°F) (Figures 5.8 and 5.9, respectively) showed more even surfaces covered only to a limited extent by graphite nodules.

During the corrosion process, these surfaces will be covered by corrosion products such as ferrous sulfate (the sole corrosion product on mild steel), carbonaceous debris, and silicic acid derived from the oxidation of silicon dissolved in the iron.[98] These corrosion products would fill any asperities left on the surfaces by the dissolution process (cf. for example Figure 5.6). They are also very easily removed when the surfaces are cleaned for weight-loss measurements at the end of each experiment.

Thus far, the existence of surface films on the corroded cast iron specimens has been proved by SEM. To carry the analysis further, the value of the mass transfer factor through the corrosion product K_{FILM} is calculated from Equation (5.17). Before that, K_{CALC} must be determined from the Eisenberg relation as shown in Section 4.3 and K_{MEAS} calculated from Equation (5.16). The values obtained for K_{FILM} at 40°C (104°F) are listed for all the alloys in Appendix D, Tables D.30 to D.33. The values reported have been calculated from the measured corrosion current at the end of the exposure time at each rotation velocity. Figure 5.4 (A and B) is a plot of K_{MEAS}, K_{CALC}, and K_{FILM} against rotation velocity at 40°C (104°F). Similarly, Table D.35 of Appendix D lists the values of the different mass transfer factors as well as the calculated P values at 27°C (81°F), while Figure 5.5 (A and B) is a plot of K_{MEAS}, K_{CALC}, and K_{FILM} against rotation velocity at that temperature.

From the calculated value of K_{FILM} (Equation 5.17), it is, in principle, possible to estimate e, the thickness of the corrosion product layer, using the equation $K_{FILM} = Df/e$. At this stage, this equation is used as a convenient notation because it provides a simple physical insight into mass transfer and also because of its simple mathematical formulation. It says that the mass transfer factor is proportional to the diffusion coefficient and inversely proportional to the unknown film thickness, e. The mass transfer factor still varies in some unknown fashion with variables such as the rotational velocity (or fluid velocity) and film properties because these variations are lumped together in the unknown film thickness, e. Also, the value for Df, the diffusion coefficient of ferrous ion through the corrosion products, must be known. In the absence of such data, it is sufficient to assume that the diffusion coefficient, Df, is of the same order of magnitude for all alloys at all conditions tested. Such a value (or range of values) would, of course, be lower than the diffusion coefficient, D, of ferrous ion in sulfuric acid. In this situation, higher values of e should be expected at the lower values of K_{FILM}, i.e., for the alloys showing the least corrosion currents. (cf. Appendix D, Tables D.30 through D.33 and D.35, and Figures 5.4B and 5.5B). This is again quite obvious from the SEM micrographs produced.

The corroded surface of Mondi B at 27°C (81°F) (Figure 5.10) has a more prominent peak and valley structure than either Mondi A (Figure 5.8) or ductile iron (Figure 5.9) at the same temperature. In this situation, it is expected to retain more corrosion products and to show lower corrosion currents and flow sensitivity as indeed observed (cf. Chapter Four and Figures 4.8, 4.5, and 4.6 for the respective current variations with time and rotation velocity). However, the difference between the corroded Mondi B surfaces at 27°C (81°F) and 40°C (104°F) is more difficult to evaluate. Both structures (Figures 5.10 and 5.11, respectively) show the peak and valley aspect already mentioned together with a high-surface coverage by the graphite nodules. At both temperatures, the alloy showed largely reduced corrosion currents and loss of flow sensitivity (cf. Figures 4.8 and 4.13).

The occluded surface structure of grey cast iron is also probably more capable (Figures 5.6 and 5.7) of retaining the corrosion products in the surface asperities. This would lead to the lower currents and the complete loss of flow sensitivity observed for this alloy (cf. Sections 4.4.4 and 4.3.4, and Figures 4.12 and 4.7 for the corrosion current variations). It is equally likely that the diffusion coefficient, Df, through such a structure, because of its tortuosity, is lower than for the other alloys.

The variable proportion of corrosion products retained on the surface as well as the surface coverage by graphite nodules or flakes will determine:

- the degree of corrosion current reduction,
- the extent of corrosion current flow sensitivity, and

- the noble shift in corrosion potential observed.

All three effects are closely related because of their common origin; namely, the nature of the film.

The attributes of the film that are responsible for the observed corrosion behavior must have their origin in the cast iron analysis and morphology. These properties would include:

- extent of surface coverage by nodules or flakes,
- depth of retained corrosion products, and
- tortuosity of the diffusion path through the corrosion products. This is particularly true of grey cast iron.

Examination of the chemical analysis revealed significant differences in the magnesium content (Table 3.1) of the cast irons under investigation. Magnesium is usually present combined with the sulfur as magnesium sulfide inclusions. These have been known to increase the corrosion of modular cast irons as compared to grey cast irons.[112] The distribution of the inclusions, whether they are clustered or finely dispersed, is also important. The work of Berendson and Wranglen revealed that clustered magnesium sulfide inclusions lead to higher corrosion rates.[112] Unfortunately, and due to time limitations as well as being outside the scope of the present work, this point could not be investigated. Nevertheless, the variation of X, the ratio of measured corrosion current to calculated current, vs magnesium content at 27°C (81°F) and 40°C (104°F) for the four alloys investigated has been examined. All the X values plotted in Figure 5.12 are those obtained at the end of the exposure time at 2 ms^{-1}, as reported in Appendix D, Tables D.30 through D.33 at 40°C (104°F) and Tables D.20, D.23, D.26, and D.29 at 27°C (81°F). Since at 27°C (81°F) the two working electrodes used (except Mondi B) showed different corrosion currents, two values for X were found.

The points obtained at 40°C (104°F) indicated that grey cast iron, which has the least magnesium content of 0.006%, also shows the lowest X ratio. Mondi B at 0.028% magnesium content has the next higher ratio, and ductile iron, with the largest magnesium content of 0.042%, has a still higher current ratio. In these conditions, one could then surmise that a magnesium content of about 0.030% (Mondi A has 0.032% Mg) is the limit above which corrodibility of the matrix is increased.

In the work of Berendson and Wranglen, the authors suggest that in the case of increased corrosion of nodular irons due to magnesium sulfide inclusions, a remedy could be an improved production method in order to achieve a proper separation of magnesium sulfide from the melt before casting.[112] The importance of this aspect is mentioned in *Metals Handbook*.[113] Another improvement could be achieved by replacing magnesium with cerium since cerium also facilitates the formation of spheroidal graphite. Practical experience of cerium in steel metallurgy has not so far indicated any detrimental effect on the corrosion process.[114]

Any possible effect of nodule size on the corrosion behavior of the nodular irons, Mondi B, Mondi A, and ductile iron, is also unclear.

Coarse nodules are known to be less strong and easier to remove under the combined action of flow and electrochemical dissolution than fine nodules.[98] There is no evidence of such an effect here since there seems to be a distribution of different nodule sizes on all spheroidal irons studied (cf. Figures 5.8, 5.10, and 5.11); however, ductile iron showed finer nodules (cf. Figure 5.9). Because of the supposedly counteracting effects on corrosion of a small nodule size and a high magnesium content, it stems from the behavior of ductile iron (cf. Figure 5.12) that the detrimental influence of the latter will be predominating over any beneficial effect of the former.

Attempts at linking the corrosion behavior to the carbon content of the different cast irons did not show any noticeable pattern. As a consequence, this was not pursued further.

Due to the general scope of the present work towards generating electrochemical data under hydrodynamically controlled conditions, less emphasis could be put on investigating aspects of microstructure and chemical analysis of the cast irons studied.

In effect, these conditions turned out to be of utmost importance to elucidate a corrosion mechanism where, due to the presence of surface films, the two main aspects are:

- current reduction below the values predicted by a convective diffusion model, and
- loss of corrosion current flow sensitivity to a varying degree.

The features of the alloy in general (nodule size and shape, magnesium content and distribution) and of the film in particular (thickness, tortuosity, surface coverage of nodules or flakes), recognized in a speculative manner to be significant in this study, go some way to explain the range of corrosion behavior observed. However, more work is still needed, particularly in:

- estimating the diffusion coefficient, Df, through the retained corrosion products and comparing film thicknesses calculated analytically with the value measured experimentally (this has been done by Mahato et al. for the corrosion of steel pipe in natural flowing waters),[17]
- studying the effect of magnesium content and of its distribution as magnesium sulfide inclusions, and [112]
- finding ways to accurately estimate surface coverage of nodules or flakes and correlating it with the observed corrosion behavior.

In the next section, a summary of the proposed mechanism for the corrosion of cast irons in concentrated sulfuric acid is presented. It is also shown that a deeper physical insight into the film behavior can be gained from a more detailed treatment of the diffusion characteristics of the reacting species across the thickness of the film. This may help to better categorize the nature and behavior of the films formed on the different cast irons. It is also an essential step towards using the present results for the prediction of hydrodynamic corrosion of cast irons in concentrated sulfuric acid.

5.2.5 Summary of the Proposed Mechanism for the Corrosion of Cast Irons: The Diffusion Properties of the Surface Films

From the aspects of the dissolution mechanism of cast irons discussed so far, a model for this type of corrosion emerges. While the convective diffusion model for the corrosion of mild steel in concentrated sulfuric acid assumes that the surface is covered with a layer of ferrous sulfate (which remains constant with time, and any loss of which is made up by corrosion of the metal), the situation on cast iron surfaces is considerably less simple.[12] The reported complications are:

- The corrosion product is still mainly ferrous sulfate, though carbonaceous debris, silicic acid, and graphite nodules or flakes are also present.[98]
- The thickness of the retained corrosion-products layer is probably increasing with time and temperature. This appears to be particularly true of grey cast iron and Mondi B (cf. Section 5.2.4).
- Diffusion through the corrosion products is, in general, complicated by the morphology of the film, i.e., this is particularly the case on grey cast iron because of the tortuosity of the film.

Initially, the corrosion rates of freshly prepared cast iron surfaces are very high and particularly so at 40°C (104°F) (Figures 4.10 to 4.13). They are also, at that time, independent of the cast iron type. As corrosion proceeds, the surfaces start to become less uniform as the iron-rich areas will corrode faster than the more noble areas such as graphite nodules or flakes. The result is the observed peak and valley aspect of the nodular iron surfaces (cf. Figures 5.10 and 5.11) and the occluded cell structure on grey cast iron (cf. Figures 5.6 and 5.7). On such surfaces, the corrosion products (ferrous sulfate, carbonaceous debris, silicic acid, etc.) can and will accumulate. There will also be a varying degree of surface coverage by graphite nodules and flakes. Both effects lead to a gradual decrease of corrosion current and to a partial-to-total loss of corrosion-current flow sensitivity as the rate determining step of corrosion is no longer diffusion through the liquid concentration boundary layer but diffusion through the corrosion products film. This change is observed to have a significant influence on corrosion behavior because diffusion through the corrosion products layer is slower than that through the liquid boundary layer. The ratio, $\frac{Df}{D}$, will be less than unity and will also probably vary with the nature of the cast iron (cf. Section 5.2.4.).

The effect of temperature is relatively simple to unravel. The initially high-corrosion currents observed at 40°C (104°F) will cause accentuation (i.e., deepening) of the characteristic peak-and-valley and occluded cell structure observed on the cast iron surfaces (cf. for example, Figures 5.6 and 5.7).

Accumulation of corrosion products on the surface will be greater and will constitute a higher resistance to mass transfer at 40°C (104°F) than at 27°C (81°F) (cf, Figures 5.4[B] and 5.5[B] and also Tables D.30 through D.33 of Appendix D). The overall effect will be the very low measured-to-calculated current ratios observed, particularly on grey cast iron and Mondi B, and the virtually complete loss of flow sensitivity for all the alloys at that temperature (cf. Tables D.30 through D.33 of Appendix D).

The model proposed has identified K_{FILM}, the corrosion products film, as the added impediment to mass transport leading to the observed corrosion behavior. The film mass-transfer factor has also been found to decrease with velocity. This is an unexpected result that says that the film will constitute a higher resistance to mass transfer the higher the velocity. In case of an unchanged corrosion current upon increase in velocity, the calculated decrease in the film factor (cf. Section 5.2.4) would suppose either a lower diffusion coefficient in the film or a thicker film at the higher velocities. This and other questions that are essential to the behavior of the film under flowing conditions, and thus to the overall corrosion behavior, are assessed below.

When fresh cast iron surfaces are immersed in concentrated sulfuric acid, and before any film starts to form, the corrosion currents are observed to vary significantly with flow variations. The liquid boundary layer constitutes the major resistance to mass transfer, and as it becomes thinner at higher velocities, the corrosion currents are observed to increase (cf. Section 5.2).

At later stages of the corrosion process, a film of corrosion products starts to accumulate on the surfaces, and, depending on the particular morphology of the cast iron and temperature (cf. Section 5.2.4), this film will more or less impede diffusion of the reacting species from the metal-film interface to the film-liquid interface. At sufficiently long times, the currents are observed to decrease very slowly. If one assumes that steady state film growth has been reached, then by considering a parallelopiped in the film at the very interface of the film-liquid boundary (cf. Figure 5.13), and equating the molar flux of the reacting species, j_{FILM}, into the parallelopiped to the flux j_s out, one obtains:

$$j_{FILM} = j_s \tag{5.18a}$$

$$K_{FILM}(c_s - c_i) = K_{CALC}(c_i - c_b) \tag{5.18b}$$

This also gives, by rearrangement, the interfacial concentration, c_i, at the film-liquid boundary:

$$c_i = \frac{K_{FILM}c_s + K_{CALC}c_b}{K_{FILM} + K_{CALC}} \tag{5.19}$$

The interfacial concentration, c_i, will, in the general case, assume a value somewhere between c_b and c_s, and would in the limiting cases of:

(a) a film of no resistance: i.e., $K_{FILM} >> K_{CALC}$
have a value: $c_i = c_s$

(b) a film of very high resistance: i.e., $K_{FILM} << K_{CALC}$
have a value: $c_i = c_b$

The two situations are depicted in Figure 5.13 by the concentration profiles (1) and (2), respectively. In case (a), the film is behaving as if it was not present; all the resistance to mass transport is in the liquid boundary layer. In case (b), the film is of such a high resistance that the resistance through the liquid layer appears to have vanished.

The question of interest now is: What are the actual resistances afforded by the films formed on the different cast irons with respect to the two extreme cases depicted above?

If the film, because of its morphology and thickness, does not constitute a significant resistance to mass transport, no large concentration gradient of the diffusing species is allowed to develop across its thickness. This would, for example, be the case of Mondi A and ductile iron at 27°C (81°F) (cf. Section 5.2.4). As a consequence, the concentration profile will be similar to the case of a film of no resistance idealized by curve (1) on Figure 5.13, and the value of K_{FILM}, the film mass transfer factor, will be high compared to the value of K_{CALC} predicted by the convective diffusion model.

Under steady state conditions, the flux will be equal to j_s (Equations 5.18a and 5.18b) and

$$j_s = K_{CALC}(c_i - c_b) \tag{5.20}$$

By replacing c_i with its expression in Equation (5.19), one is able to write:

$$j_s = \frac{K_{CALC}K_{FILM}}{K_{CALC}+K_{FILM}}(c_s - c_b) \tag{5.21}$$

and if $K_{FILM} >> K_{CALC}$, Equation (5.21) can be reduced to:

$$j_s \simeq K_{CALC}(c_s - c_b) \tag{5.22}$$

Equation (5.22) simply states that if the film mass transfer factor is highly compared to the mass transfer factor calculated from the convective diffusion model, then the measured current can adequately be described by the equation for convective diffusion.

If, however, the film significantly impedes mass transport as with, for example, grey cast iron (cf. Section 5.2.4), a substantial concentration gradient will develop across the thickness of the film. This will be similar to the case of a film of very high resistance idealized by curve (2) in Figure 5.13. The value of K_{FILM} will be lower than, or similar to, the value of K_{CALC} predicted. For example, when $K_{FILM} = \frac{K_{CALC}}{3}$,

$$j_s = j_{FILM} = K_{FILM}(c_s - c_i) \tag{5.23a}$$

combining with Equation (5.19)

$$j_s = j_{FILM} = \frac{K_{CALC}}{4}(c_s - c_b) \tag{5.23b}$$

Thus, when $K_{FILM} = \frac{K_{CALC}}{3}$, the measured current would be expected to be only one quarter of the value predicted by the model. When the film diffusion properties are close to the one depicted by curve (1), the major resistance to mass transport is in the liquid boundary layer, and the corrosion rate is expected in such circumstances to show a significant sensitivity to flow variations as indeed is the case with, for example, Mondi A and ductile iron at 27°C (81°F). When the film is thicker and the surfaces are covered to an appreciable extent by graphite flakes (grey cast iron) and nodules (Mondi B), the diffusion starts to become difficult, and a concentration gradient can develop, (line [3] represents an intermediate case of a film of some resistance).

The effect of velocity on the corrosion behavior is more complex. First, flow variations have a limited effect on the corrosion currents measured. Also, the film factor is observed to decrease with velocity. This can be explained as follows: an increase in velocity will lead to a lower value of c_i, i.e., c_i gets nearer to c_b, and the concentration profile is consequently steeper. Since the experimental observation of a lower film factor at high velocities must be explained by either a lower diffusion coefficient or a thicker film, the steeper concentration profile can explain both since it necessitates either a lower diffusion coefficient or a longer effective diffusion path because of the tortuosity of the film. The concept of film tortuosity introduced in studies of liquid diffusion in porous media has been defined as:

$$D_{eff} = \frac{Df}{\alpha} \tag{5.24}$$

where α, the tortuosity, typically has values of 3 to 6, though values as high as 10 have been measured.[115] D_{eff} is the effective diffusion coefficient through the tortuous path. Thus, one can rewrite the film mass transfer factor, K_{FILM}, as:

$$K_{FILM} = \frac{Df}{\alpha\theta} \tag{5.25}$$

where α, the tortuosity, must be a function of morphology and flow velocity.

The decrease of K_{FILM} with velocity has also been observed on the alloys showing a behavior close to case (1), but because of the very high values of K_{FILM}, the situation there is quite different, and the film cannot play a major role in determining the overall mass transfer characteristics at the surface.

A simple analogy with the present diffusion situation is the electric one. In case (1), where $c_i = c_s$, the film is behaving like a short circuit. However, in case (2), the tortuosity of the film will make it behave like a number of resistances in series, thus allowing an increase in the total resistance.

The introduction of the film tortuosity has helped explain the apparent paradox of surface films constituting higher resistance to mass transfer at higher velocities.

5.2.6 Implications of the Present Study on Corrosion Prediction Using the RCE or Other Geometries

The aim of any laboratory investigation of the type reported in this study is twofold:

- to gain physical insight into the mechanisms of the corrosion itself (this has already been discussed), and
- to provide the corrosion engineer with methods, analytical or empirical, to allow him to make corrosion predictions in the plant geometry.

Primarily, the corrosion behavior observed at short exposure times, i.e., the initially high corrosion rates, is of no relevance to the corrosion performance expected from, for example, a cast iron pipeline section or a cast iron vessel over a service life of several years. This, of course, would suppose within reason, that the films observed here using the RCE geometry, will also develop in other geometries. The in-plant situation would be one where a steady state film, if operational conditions allow, has formed. Depending on the cast iron type, flow velocity, acid concentration, temperature, and flow geometry, these films will determine to a greater or lesser extent the subsequent behavior.

In the present study, the initially very high corrosion currents, were subsequently observed to vary only a little with time at the considerable exposure times allowed. In an ideal situation, and with hindsight gained from analyzing the results, a run using a fresh sample each time would have been performed at each singular velocity for a long enough time, probably one to a few weeks, to reach a completely steady state corrosion. In this way, the film mass transfer factors would be more representative of the desired steady state. However, in the laboratory environment, the situation is not ideal, but it can be said with confidence that any improvement from longer or different experiments would not affect the trends in the behavior of the cast irons used in this study. The film mass transfer properties deduced from the corrosion data can then be considered to be representative of a steady state film, at least for the exposure times studied.

This in turn constitutes a reasonable ground for the generalization of the corrosion behavior observed to give a conservative estimate of the cast irons' performance at even longer exposure times.

Obviously, the next step here is to categorize the film mass-transfer characteristics, and if possible, come up with criteria that will allow the use of the present results in corrosion prediction.

A summary of the mass transfer properties of the films formed at 40°C (104°F) on all cast irons tested is pictured in Figure 5.14. At this temperature, high corrosion currents of a transient nature were observed in the first 24 h exposure. By the time the currents fell below the values predicted from the convective diffusion model, sensitivity of those currents to flow variations was already seriously reduced, indicating that a film had formed on all the alloys. It is also suggestive that the films found at 40°C (104°F) were able to give a better protection to the underlying metal surface than those formed at 27°C (81°F) when significant differences between the alloys were uncovered. The transient nature of the initially high currents makes any data obtained at short exposure times (probably less than 48 h) of little relevance to the corrosion engineer who is interested in the long-term corrosion performance of the cast iron equipment being monitored. The data obtained at 40°C (104°F) for the exposure times studied were used to calculate the film mass-transfer factor. The values of K_{FILM} derived in this manner were then plotted against the Re number on a log-log scale and reported in Figure 5.14. The points obtained for each alloy were found to lie on a straight line, and thus a single expression, in terms of the Re number and numerically derived constants, was obtained for each alloy. The expressions where K_{FILM} has units of cm/s are as follows:

Grey cast iron $K_{FILM} = 1.27 \cdot 10^6 \, Re^{-2.52}$ −18% (5.26)

Mondi B $K_{FILM} = 8.35 \cdot 10^3 \, Re^{-1.94}$ +16% (5.27)

Mondi A $K_{FILM} = 8.81 \cdot 10^2 \, Re^{-1.63}$ −13% (5.28)

Ductile iron $K_{FILM} = 8.37 \cdot 10^5 \, Re^{-2.39}$ + 6% (5.29)

The percentages given beside each expression give the maximum deviation of the data around the correlations obtained.

The differences in the expressions reported for the series of alloys studied is an analytical account, within experimental error, of the differences in behavior of the alloys studied. These differences have been first reported in an empirical manner by way of the filming factor (cf. Figures 5.1 and 5.2).

As acid was replenished at 27 h for ductile iron at 2 ms^{-1} and 40°C (104°F), there is a jump in the K_{FILM} values at 27 h and the subsequent ones. The value shown in Figure 5.14 has been calculated at 36 h, and it can be seen that it falls in line with the K_{FILM} values at 0.81 ms^{-1} and 1.62 ms^{-1}, thus allowing Equation (5.29) to be deduced.

At 27°C (81°F), two alloys (specifically, Mondi A and ductile iron) did not show a significant loss of corrosion current sensitivity to flow variation, nor a considerable deduction in corrosion currents. Neither effects can be explained by films whose behaviors can then be categorized by expressions of the types already derived at 40°C (104°F). Figure 5.15, a plot of K_{FILM} vs Re number at 27°C (81°F) and for the exposure times studied, shows that the film factor values are mostly high and, thus, are not expected to have significant influence on the overall corrosion behavior (cf. Section 5.2.5). It could be concluded then that for those two alloys, the corrosion behavior is only marginally affected by the surface films, and the diffusion model as described by the Eisenberg relation will still be adequate in predicting their corrosion behavior at 27°C (81°F).

Of the two grey cast iron working electrodes used, one showed very low corrosion currents and almost complete loss of sensitivity to flow variations. The other electrode, while showing reduced currents, conserved some sensitivity to flow variations (cf. Section 4.3.4). The reason for this dual behavior is not clear, particularly when the films formed on the two electrodes did not show any significant differences. However, it is suspected that if longer exposure times were allowed, the difference between the two electrodes would have been reduced further with the common limit being the lower current (i.e.,the lower film factor) electrode. Close examination of Tables D.24 through D.26 of Appendix D and of Figure 4.7 reveals this to be indeed the trend in the data reported. Nevertheless, and for both electrodes, the values of the surface film (K_{FILM}) plotted against the Re number fell on straight lines. An expression for the film mass transfer factor in terms of the Re number could then be deduced for each electrode. The expressions obtained for K_{FILM} in units of cm/s are as follows:

Working Electrode 1 $K_{FILM} = 6.29 \cdot 10^{-3}\ Re^{-0.39}$ **–6%** (5.30)

Working Electrode 2 $K_{FILM} = 8.9 \cdot 10^{-3}\ Re^{-0.32}$ **–7%** (5.31)

The expressions are valid for the exposure times and the velocity range studied. The percentages given beside each expression give the maximum deviation of the data around the correlation obtained.

The last alloy in the present series, specifically Mondi B, showed at the 1.62 ms^{-1} and 2 ms^{-1} velocities very low currents and a considerable loss of corrosion current sensitivity to flow variation. This was not so at 0.81 ms^{-1} where the currents measured after 88 h exposure time were very close, within experimental error, to the values predicted by the convective diffusion model (cf. Section 4.3.5). This means that a K_{FILM} value cannot be calculated at the lower velocity value. However, some extrapolation is possible from the values of K_{FILM} obtained at the higher velocities.

On Figure 5.15, a point at approximately 10^{-3} cm/s and 0.81 ms^{-1} would fall on a line passing through the three points thus obtained. The values for K_{MEAS} at 0.81 ms^{-1} derived in this fashion (after a hypothetical velocity change at the end of the 2 ms^{-1} run) would be $1.312 \cdot 10^{-4}$ cm/s. Since $c_s - c_b = 1.428 \cdot 10^{-6}$ mols/cm^3, the corrosion current corresponding to this value would be 0.352 mA. This current, higher than the value of 0.315 mA found at 2 ms^{-1} (cf. Table D.29 of Appendix D), is certainly unlikely in view of the loss of sensitivity to flow variations observed upon the velocity change from 1.62 ms^{-1} to 2 ms^{-1} (cf. Table D.28 of Appendix D and Section 4.3.5). The film that has formed at 1.62

ms^{-1} and 2 ms^{-1}, and that is responsible for the actual observed behavior at those velocities, would be expected to influence any subsequent run at 0.81 ms^{-1}. As a consequence, the measured current at 0.81 ms^{-1} will be expected to be the same or lower than the value found at 2 ms^{-1} rather than higher. If the current thus measured were 10% lower than the 2 ms^{-1} value, the K_{FILM} value would be $3.50 \cdot 10^{-4}$ cm/s, or if it remained unchanged, the K_{FILM} value would have been $5.24 \cdot 10^{-4}$ cm/s. This last point is plotted in Figure 5.15. One can see that it is somehow difficult to draw a straight line through the points thus obtained. The conclusion must be that because of the initial behavior at 0.81 ms^{-1} (cf. Table D.27 of Appendix D), no analytical expression for the film factor can be thus proposed. However, it can be concluded that the film does sufficiently impede mass transport to significantly control the corrosion behavior at the exposure times studied. This conclusion cannot be improved on by further experimentation as no fresh samples are available at this time.

In hydrodynamic corrosion research, the foremost consideration for accurate prediction of corrosion performance is the ability to transfer corrosion rates measured in one geometry to another geometry. The question has been extensively reviewed in Section 5.1.6. with particular emphasis on mild steel corrosion in concentrated sulfuric acid. Most investigators have limited the approach to corrosion rate transfer around simple geometries such as the RCE, the pipe flow, and the annular flow channel.[20,75,109] This is, of course, due to the importance of the circular pipe geometry that is frequently encountered in industrial applications. Also, in situations where more complex shapes are involved, the added difficulties may prevent a solution from being found. These are, namely, that the mass transfer characteristics of the particular geometry (i.e., pump impeller) may not be known, the enhancement of mass transfer may occur at elbows (cf. Section 5.1.6.4), and erosion may be possible.

In the cast iron situation, the film diffusion characteristics, found to control steady state corrosion to a greater or lesser extent, appear to be a function of the flow velocity and also primarily of the morphology of the corrosion products' surface film. There is every possibility that such films would not be significantly affected by the geometry. A similar result has been found in some recent work on the corrosion of a Cu-Ni alloy in aerated flowing 1 M NaCl solutions using three geometries; namely, the RCE, the pipe flow, and the annular flow channel.[75] In that particular instance, Chen et al. have found that the overall mass transfer rate can be written as the sum of the mass transfer rate in the diffusion boundary layer as predicted from previously derived mass transfer correlations (i.e., the Eisenberg relation for the RCE,[46] the Harriott and Hamilton relation for the pipe flow,[68] and the Ross and Wragg relation for the annular flow channel[116]) and of the mass transfer rate in the surface film.[75] The relations in terms of the Re number found for the films deposited in the different geometries were very similar; they were only slightly different by the multiplication constant. As a consequence, the authors were able to produce mass transfer correlations based on the overall mass transfer factor that enabled transfer of corrosion rates measured in the RCE to the annular flow channel and the pipe flow geometries. It was also proved that the corrosion rate, when plotted as a function of the overall mass transfer factor (on a log-log scale), was independent of the geometry involved.[75]

This is not unlike the present situation, and it is suspected that, depending on the alloy used and once a substantial film in terms of thickness and morphology has formed, its characteristics will (particularly at the higher temperatures and velocities) determine to a large extent any subsequent corrosion so that the geometry characteristics in terms of mass transfer become secondary. This last point implicitly supposes that conditions leading to a steady state film formation and growth are verified. Such conditions may be disturbed by very high velocities where mass transfer enhancement in the film may again lead to increased corrosion rates and a more prominent effect of the geometry. A further indication that the geometry may become unimportant when the film is capable of controlling the corrosion is that film dominance is accompanied in this study by a substantial loss of the corrosion current sensitivity to flow variations. Of course, these ideas are only hypotheses at this time and may need further experimental verification for the cast irons-sulfuric acid system in other geometrical arrangements than the RCE.

In cases where the film does not have a significant effect on the observed corrosion, the mass transfer relation pertaining to the particular geometry, if known, may still be used to predict the corrosion rate. This was found to be the case for Mondi A and ductile iron at 27°C (81°F) and over the velocity range used in this study.

Table 5.1

Range of Property Values Studied in the Present Work

Property / Acid Concentration	$\nu(\cdot 10^{-6} m^2/s)$		$D(\cdot 10^{-12} m^2/s)$		$Sc = \frac{\nu}{D}$		Re power Exponent	
	27°C (81°F)	40°C (104°F)	27°C (81°F)	40°C (104°F)	27°C (81°F)	40°C (104°F)	27°C (81°F)	40°C (104°F)
68%	4.764	3.468	66.30	96.30	71,868	36,018	0.7	0.7
98%	10.66	7.58	25.85	38.24	412,440	198,220	0.77	0.81
93%	10.0	7.11	27.61	40.76	362,009	174,456	0.77	0.81

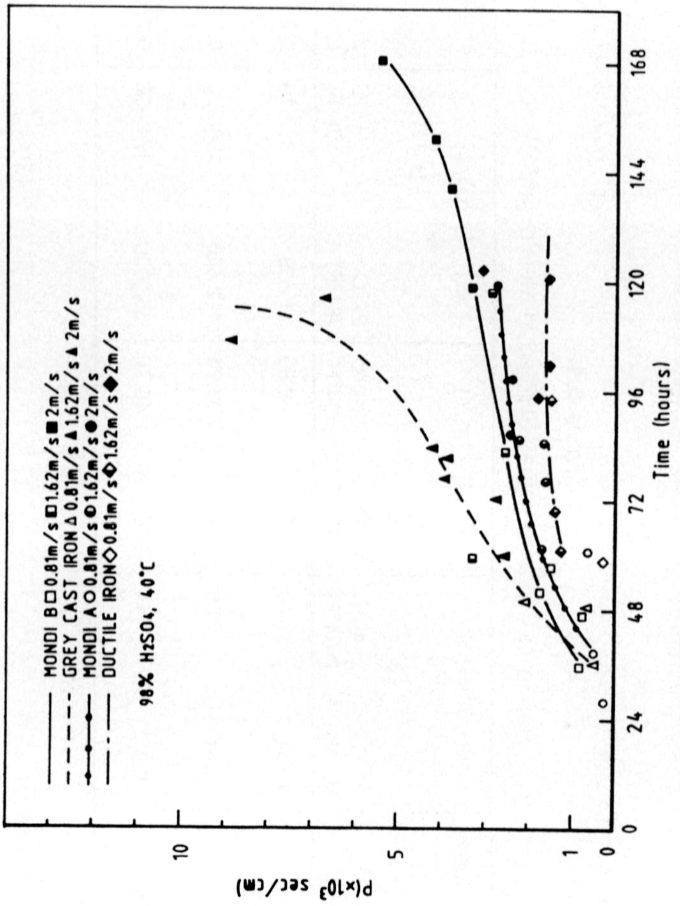

Figure 5.1 P, the filming factor, vs time for all four cast irons tested in 98% H_2SO_4 at 40°C (104°F).

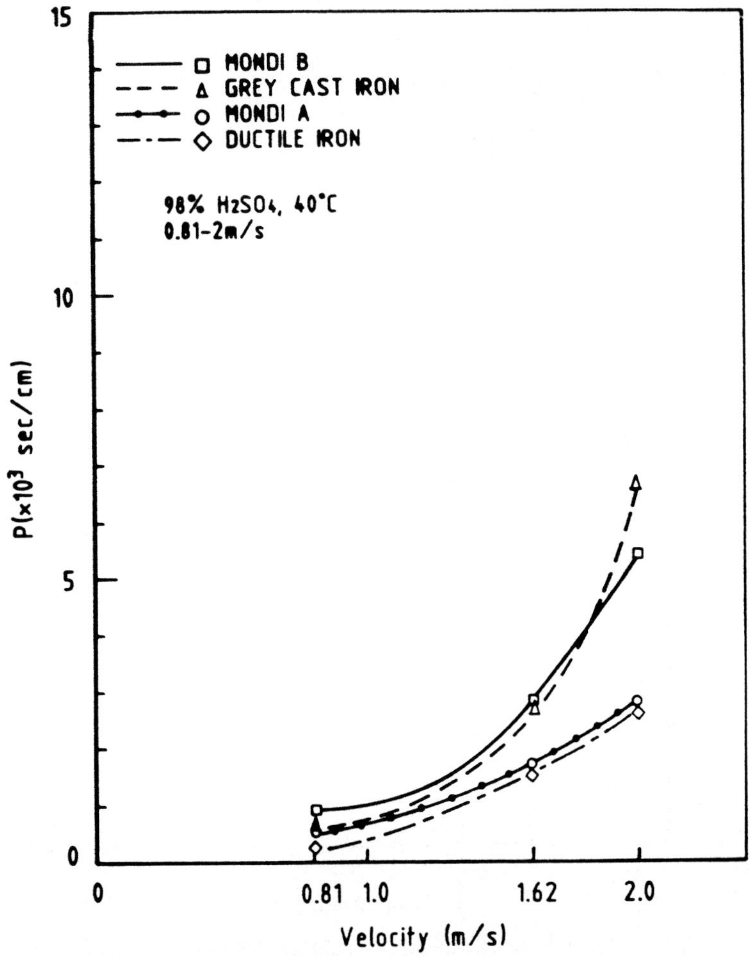

Figure 5.2 P, the filming factor, vs rotation velocity for the four cast irons tested in 98% H_2SO_4 at 40°C (104°F).

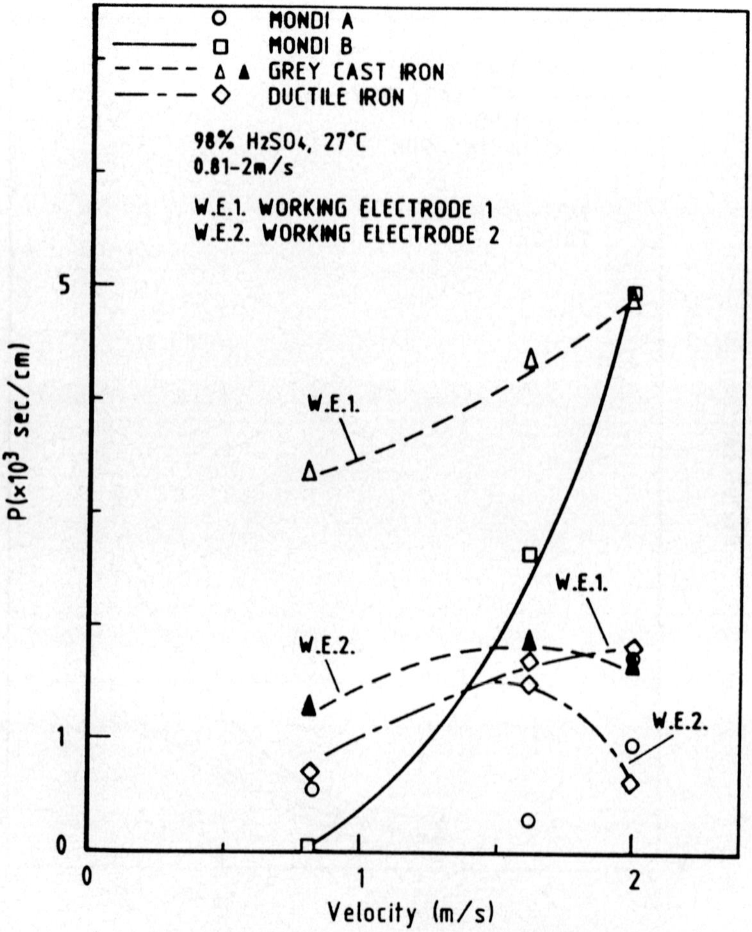

Figure 5.3 P, the filming factor, vs rotation velocity for the four cast irons tested in 98% H_2SO_4 at 27°C (81°F).

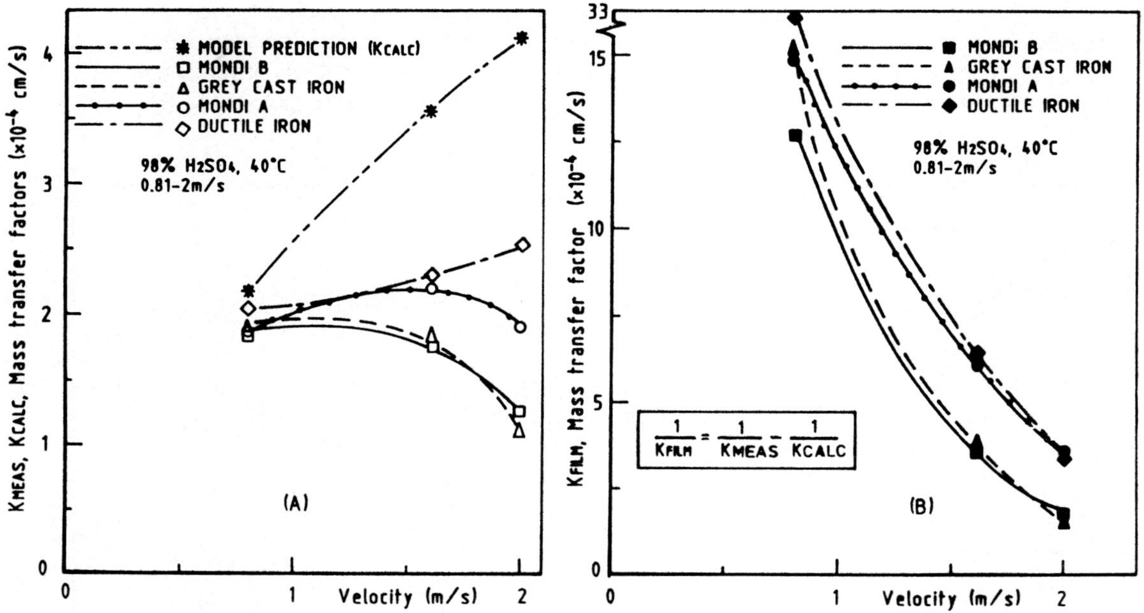

Figure 5.4 (A) Predicted mass transfer factor, K_{CALC}, and measured mass transfer factors, K_{MEAS}, vs rotation velocity for the four cast irons tested in 98% H_2SO_4 at 40°C (104°F). (B) Mass transfer factors through the corrosion product film, K_{FILM}, vs rotation velocity for the four cast irons tested in 98% H_2SO_4 at 40°C (104°F).

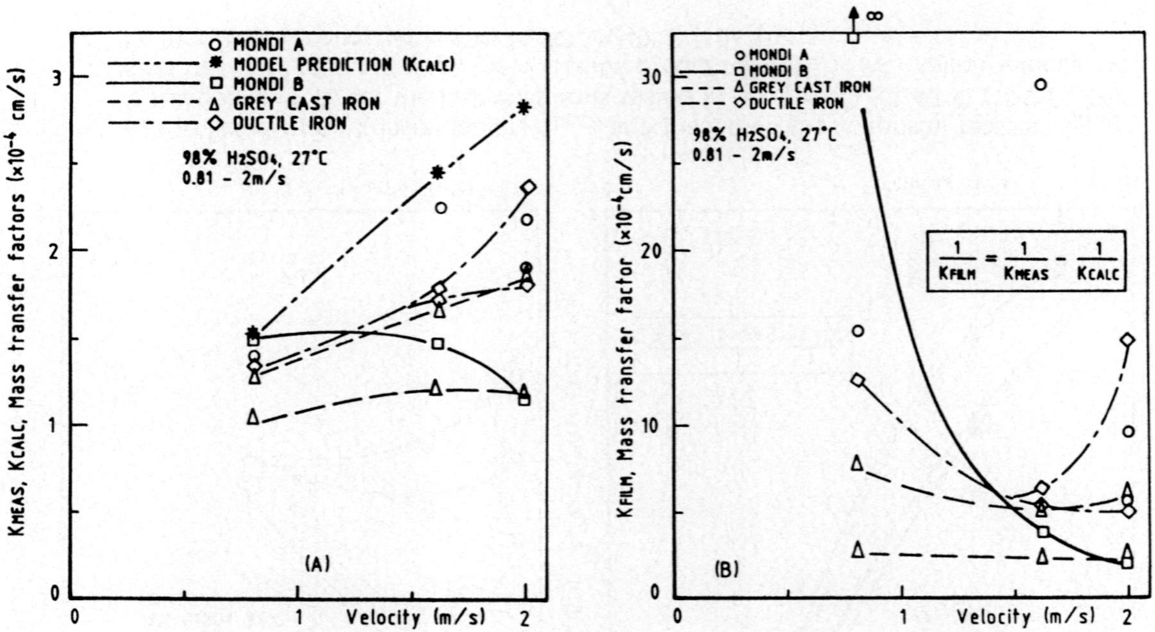

Figure 5.5 (A) Predicted mass transfer factor, K_{CALC}, and measured mass transfer factors, K_{MEAS}, vs rotation velocity for the four cast irons tested in 98% H_2SO_4 at 27°C (81°F). (B) Mass transfer factors through the corrosion product film, K_{FILM}, vs rotation velocity for the four cast irons tested in 98% H_2SO_4 at 27°C (81°F).

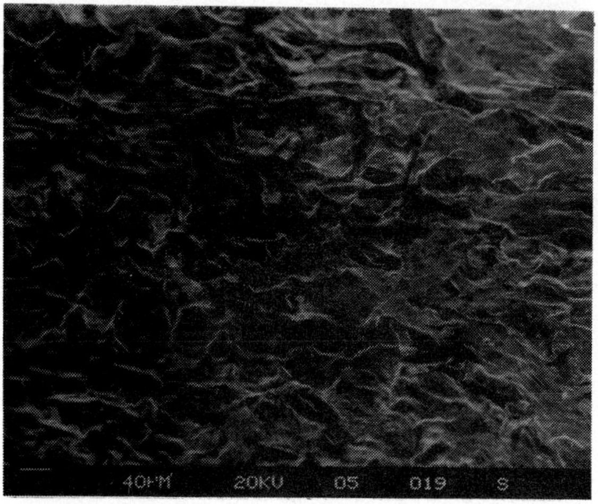

Figure 5.6 SEM micrograph of grey cast iron surface exposed to 98% sulfuric acid at 40°C (104°F).

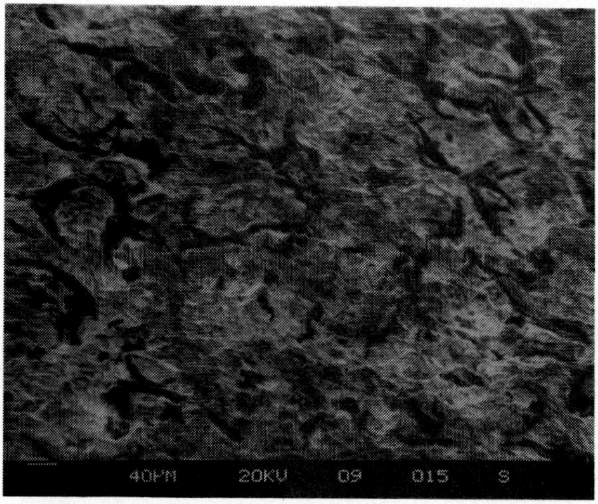

Figure 5.7 SEM micrograph of grey cast iron surface exposed to 98% sulfuric acid at 27°C (81°F).

Figure 5.8 SEM micrograph of Mondi A surface exposed to 98% sulfuric acid at 27°C (81°F).

Figure 5.9 SEM micrograph of ductile iron surface exposed to 98% sulfuric acid at 27°C (81°F).

Figure 5.10 SEM micrograph of Mondi B surface exposed to 98% sulfuric acid at 27°C (81°F).

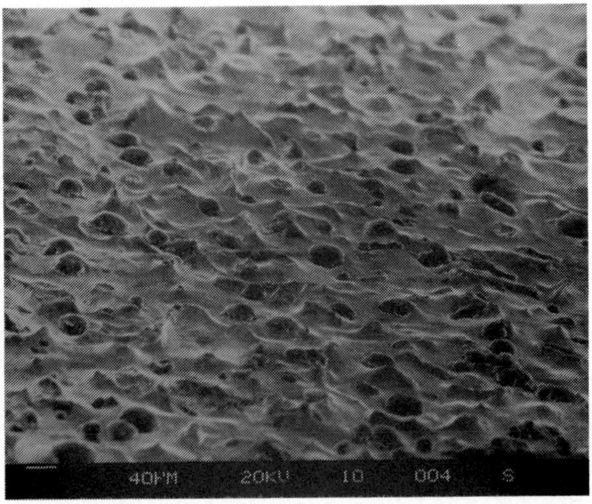

Figure 5.11 SEM micrograph of Mondi B surface exposed to 98% sulfuric acid at 40°C (104°F).

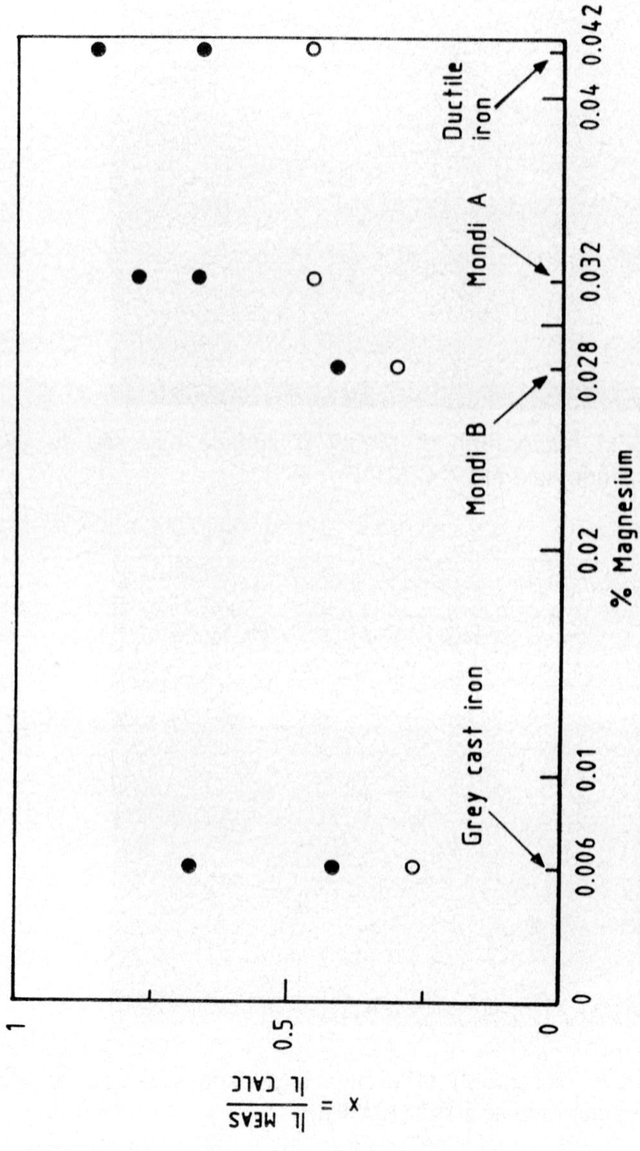

Figure 5.12 Variation of measured-to-calculated corrosion current ratio vs magnesium content at 40°C (104°F) (O) and 27°C (81°F) (●) for all cast irons in 98% H_2SO_4. Data reported is at end of exposure at $2 ms^{-1}$.

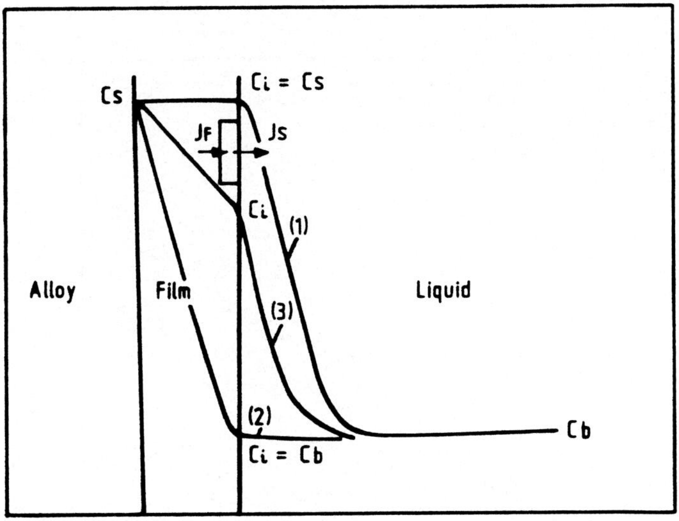

Figure 5.13 Schematic diagram of the different concentration profiles across the films formed on the cast iron surfaces.

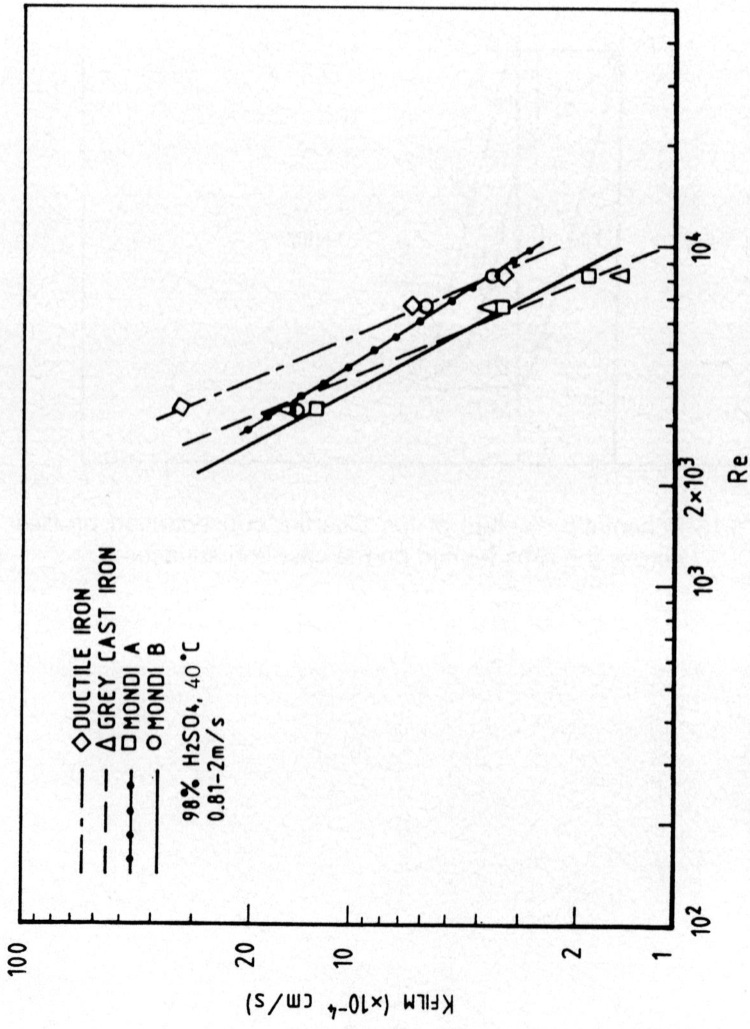

Figure 5.14 Film mass transfer factor, K_{FILM}, vs Re for all cast irons tested at 40°C (104°F).

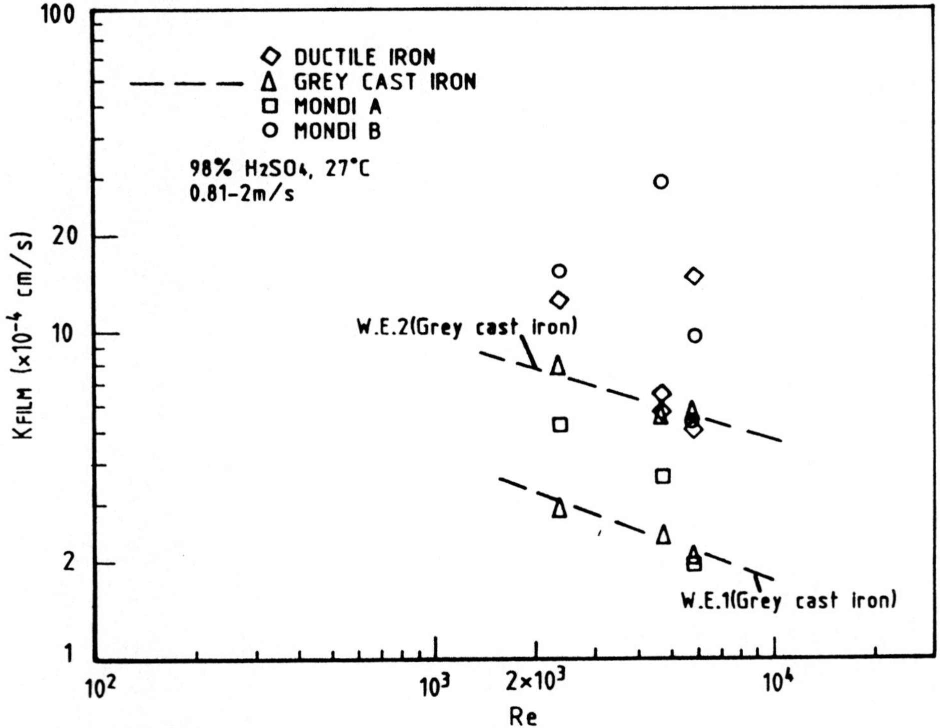

Figure 5.15 Film mass transfer factor, K_{FILM}, vs Re for all cast irons tested at 27°C (81°F).

CHAPTER
SIX

CONCLUSIONS

- A previous investigation of the corrosion of mild steel in 68% sulfuric acid using the rotating cylinder electrode (RCE) has shown that the corrosion rate is proportional to the velocity raised to the power of 0.7. It has also been suggested that this may be the case for the 50% to 99% acid concentration range. The present study, having confirmed the findings at the 68% acid concentration level, has uncovered a significant variation of the velocity exponent with acid concentration, as well as with temperature.

- The present results have shown the velocity dependence exponent to be 0.77 in 93% and 98% sulfuric acid at 27°C (81°F) and 0.81 in 93% and 98% sulfuric acid at 40°C (104°F), thus leading to the significant increase in corrosion currents observed. Nevertheless, the sensitivity of the corrosion currents to flow variations and to bulk acid ferrous ion content suggested that the convective diffusion of ferrous ion away from a saturated ferrous sulfate film-liquid interface into the bulk of the acid was still controlling the corrosion rate.

- The Schmidt number (Sc) values encompassed by this study were well outside the range reported by previous investigators. The velocity exponent dependence with acid concentration and temperature has been linked to a possible influence of the physical characteristics (Sc) of the dissolution mechanism on the hydrodynamics (Reynolds number [Re] exponent) under which mass transfer occurs. In any case, the present results may also mean that the applicability of the previously derived mass transfer relation cannot be extended to Sc values much higher than those reported earlier.

- It has also been suggested that the observed increase of the mass transfer rate in 93% and 98% sulfuric acid may be due to the nature of the ferrous sulfate film. Previous research has shown that above approximately 99% acid concentration, corrosion rates of mild steel are higher than those predicted from a ferrous sulfate diffusion model. Therefore, other mechanisms are also operating in this range. There may be some crystallographic or structural changes that occur in ferrous sulfate that make it lose its adhesion to the steel. If such a transformation is possible, the present results could be considered to suggest that this happens at a lower concentration than 99%.

- The specific design features of the geometrical arrangement of the RCE used substantiated previous findings about the negligible influence of a number of geometrical variations on the mass transfer process. In this study, an auxiliary electrode mounted on the rotating shaft, rather than the usual outer concentric cylinder, was used. It has been suggested that this choice, which is dictated by practical as well as economical reasons, cannot significantly influence a mass transfer process controlled by hydrodynamic conditions near the wall. Indeed, the agreement found between this study and previous work in 68% sulfuric acid using an outer concentric-cylinder auxiliary electrode can be considered to be an excellent check for the apparatus used, giving added confidence to the results reported in this study.

- The dependence of the velocity exponent on acid concentration and temperature observed in the RCE geometry may be occurring in other geometries of more direct industrial application, particularly in the pipe flow geometry. This point and other questions of interest to the transfer of corrosion rate from one geometry to another have been discussed in the light of the present results.

- Cast irons were observed to corrode with the formation of surface films of retained corrosion products. The surface films existence and nature in terms of the surface coverage by graphite nodules or flakes, morphology, and thickness, have been confirmed and examined by Scanning Electron Microscopy (SEM).

- The presence of the surface films was shown to present an additional impediment to convective mass transfer of the reacting species. As a consequence, a model was suggested whereby the overall measured mass transfer factor is the sum of the mass transfer factor in the liquid boundary layer and the mass transfer factor in the surface film.

Using the model put forward, it was possible to explain the main experimental features observed for the cast irons studied; these are, namely, the reduction of the corrosion currents below the values expected from the convective diffusion model and the partial-to-total loss of sensitivity to flow variations of the measured corrosion currents.

- The corrosion currents reduction and the loss of sensitivity to flow variations of the corrosion currents were observed to increase with rotation velocity, temperature, and exposure time. Both effects were, in general, more pronounced for grey cast iron and Mondi B than for ductile iron and Mondi A. The velocity effect on corrosion was found to depend on the surface films' morphology in a complex manner and was recognized to affect the reacting species' diffusion profile across the surface films, thus leading to the observed effect on the corrosion currents.

- At 40°C (104°F) and for the exposure times and the velocity range studied, the mass transfer factors of the surface films formed on all cast irons studied were expressed in terms of the Re and other numerically determined constants. Analysis of the experimental results has shown that the surface films' mass transfer characteristics are expected to control the corrosion rate. Therefore, it has been suggested that knowledge of the surface films' steady state mass transfer factors is foremost for corrosion prediction when the effect of the flow geometry, RCE, or possibly some other geometry, on the overall measured mass transfer rate becomes secondary. The existence of the surface films also means that corrosion rates reported at short exposure times (less than about 48 h) are of no relevance to in-plant corrosion prediction where corrosion rates obtained over considerable exposure times are needed.

- At 27°C (81°F), two alloys (specifically, Mondi B and grey cast iron) showed behaviors similar to those observed at 40°C (104°F). The surface films formed on the two alloys are thus expected to control corrosion to a large extent. Knowledge of the surface films' mass transfer characteristics in terms of the Re will also be essential in predicting their corrosion behavior. For the other two alloys (specifically, Mondi A and ductile iron), the corrosion currents measured were, in general, close to the values expected from the convective diffusion model, and no appreciable loss of sensitivity to flow variations of the corrosion currents was experienced. It can thus be concluded for those two alloys and over the velocity range studied that the convective diffusion model can still be useful in predicting their corrosion behavior. The features of the different cast irons in terms of thickness and morphology of the surface films revealed by SEM substantiated to a reasonable extent the hydrodynamic corrosion data.

LIST OF REFERENCES

1. G.A. Nelson, *Materials Protection* 3,1(1964): pp. 24-33.

2. O.T. Fasullo, *Sulphuric Acid Use and Handling* (New York, NY: McGraw-Hill Book Company, 1965), pp. 51-64.

3. K.A. Kobe, R.E. Fredrickson, *Journal of Chemical Engineering Data* 1(1956): pp. 13-17.

4. D. Fyfe, R. Vanderland, J. Rodda, *Chemical Engineering Progress* 73(1977): pp. 65-69.

5. NACE Publication 5A151, "Materials of Construction for Handling Sulphuric Acid" (Houston, TX: NACE, 1985).

6. H.J. French, *Alloy Constructional Steels* (Metals Park, OH: American Society for Metals, 1942), pp. 182-184.

7. E. Shields, W.J. Dessert, *Pollution Engineering* 13, 12(1981): pp. 39-40.

8. S.W. Dean, G.D. Grab, "Corrosion of Carbon Steel Tanks in Concentrated Sulfüric Acid Service," CORROSION/85, paper no. 298 (Houston, TX: NACE, 1985).

9. B.T. Ellison, W.R. Schmeal, *Journal of the Electrochemical Society* 125(1978): pp. 524-531.

10. L.A. Poluboyartseva et al., *Journal of Applied Chemistry USSR* 36(1963): pp. 1,210-1,217.

11. G.H. Damon, *Industrial and Engineering Chemistry* 33(1941): pp. 67-69.

12. S.W. Dean, G.D. Grab, "Corrosion of Carbon Steel by Concentrated Sulfüric Acid," CORROSION/84, paper no. 147, (Houston, TX: NACE, 1984).

13. D.R. Gabe, D.J. Robinson, *Electrochimica Acta* 17(1972): pp. 1,129-1,137.

14. V.G. Levich, *Physiochemical Hydrodynamics* (Englewood Cliffs, NJ: Prentice Hall Inc., 1962), p. 20.

15. B. Poulson, *Corrosion Science* 23, 4(1983): pp. 391-430.

16. L.L. Shrier, ed., *Corrosion*, 1976 edition, vol.1, (London, United Kingdom: Newnes-Butterworth, 1976), p. 2:12.

17. B.K. Mahato, C.Y. Cha, L.W. Shemilt, *Corrosion Science* 20(1980): pp. 421-441.

18. K.D. Efird, *Corrosion* 33, 1(1977): pp. 3-8.

19. D.R. Gabe, *Journal of Applied Electrochemistry* 4(1974): pp. 91-108.

20. G. Wranglen, J. Berendson, G. Karlberg, "Physicochemical Hydrodynamics Conference in Honor of V. Levich," D.B. Spalding, ed. (London, United Kingdom: Advance Publications Ltd., 1977), pp.461-472.

21. W. Linton, T. Sherwood, *Industrial and Engineering Chemistry* 26(1934): p. 516.

22. T. Chilton, A.P. Colburn, *Industrial and Engineering Chemistry* 26(1934): pp. 1,183-1,187.

23. A.P. Colburn, *Transactions of the American Institute of Chemical Engineers* 29(1933): pp. 174-209.

24. E.L. Cussler, *Diffusion, Mass Transfer in Fluid Systems* (Cambridge, United Kingdom: Cambridge University Press, 1984), p. 118.

25. J.S. Newman, Electrochemical Systems, *Prentice Hall Series in the Physical and Chemical Engineering Sciences* (Englewood Cliffs, NJ: Prentice Hall Inc., 1973), pp. 301 and 305.

26. C.W. Tobias, M. Eisenberg, C.R. Wilke, *Journal of the Electrochemical Society* 99, 12(1952): pp. 359c-365c.

27. Z. Zembura, *Corrosion Science* 8(1968): pp. 703-713.

28. C.V. King, M.M. Burger, *Journal of the Electrochemical Society* 65(1934): p. 403.

29. M.B. Abramson, V. King, *Journal of the American Chemical Society* 61(1939): p. 2,290.

30. R. Glicksman, H. Mouguin, C.V. King, *Journal of the Electrochemical Society* 100(1953), p. 580.

31. M. Eisenberg, C.W. Tobias, C.R. Wilke, *Journal of the Electrochemical Society* 101, 6(1954): pp. 306-319.

32. U.R. Evans, *An Introduction to Metallic Corrosion*, 3rd edition, (London, United Kingdom: Edward Arnold, 1981), p. 85.

33. L.L. Shrier, ed., *Corrosion*, 1976 edition (London, United Kingdom: Newnes-Butterworth, 1976), p. 1:88.

34. P.O. Gartland et al., *Corrosion* 40, 3(1984): pp. 127-133.

35. T.K. Ross, G.C. Wood, I. Mahmus, *Journal of the Electrochemical Society* 113, 4(1966): pp. 334-345.

36. R.G. Hickman (Ph.D. thesis, Univ. of California, Berkeley, 1963).

37. A.A. Wragg, *The Chemical Engineer*, Jan. (1977): pp. 39-49.

38. J.R. Selman, G.W. Tobias, *Advances in Chemical Engineering*, vol. 10, T.B. Drew, et al., eds., (New York, NY: Academic Press, 1978), pp. 212-318.

39. T. Mizuchina, *Advances in Heat Transfer*, vol. 7, T.F. Irvine, J.P. Hartnett, eds. (New York, NY: Academic Press, 1971), pp. 87-161.

40. D.A. Frank-Kamenisti, *Diffusion and Heat Exchange in Chemical Kinetics*, Translated from the Russian by N. Thon (Princeton, NJ: Princeton University Press, 1955), p. 18.

41. K. Enever, (Cranfield Fluid Engineering Unit Short Course paper on Similarity, Cranfield, United Kingdom, 1985).

42. O. Reynolds, *Proceedings of the Manchester Literary and Philosphical Society*, vol.14(1874): page 7.

43. V.G. Levich, *Physiochemical Hydrodynamics* (Englewood Cliffs, NJ: Prentice Hall Inc., 1962), p. 59.

44. V.G. Levich, *Physiochemical Hydrodynamics* (Englewood Cliffs, NJ: Prentice Hall Inc., 1962), pp. 60-72.

45. J.S. Newman, *Electrochemical Systems* (Englewood Cliffs, NJ: Prentice Hall Inc., 1973), p. 307.

46. M. Eisenberg, C.W. Tobias, C.R. Wilke, *Chemical Engineering Progress Symposium Series* 51, 16(1953): pp. 1-16.

47. C. Friedengagen, *Z. Anorg. Chem.* 29(1902): p. 396.

48. G. Grube, *Zeitschrift für Elektrochemie* 18(1912): p. 189.

49. S. Ilmori, *Z. Anorg. U. Allgem. Chem.* 167(1927): p. 145.

50. C.S. Lin et al., *Industrial and Engineering Chemistry* 43(1951): p. 2,136.

51. S.L. Goren, R.V.S. Mani, *American Institute of the Chemical Engineer Journal* 14(1968): p. 57.

52. P.A. Lush et al., *Proceedings of the Sixth European Congress on Metallic Corrosion* (London, United Kingdom: 1977), pp. 137-146.

53. K.M. Krall, E.M. Sparrow, "Transactions of the American Society of Mechanical Engineers," *Journal of Heat Transfer* Feb.(1966): p. 131.

54. D.J. Tagg, M.A. Pattrick, A.A. Wragg, *Transactions of the Institute of Chemical Engineers* 67(1979): p. 176.

55. E.M. Sparrow, J.L. Gregg, "Transactions of the ASME," *Journal of Heat Transfer* 81(1959): p. 249.

56. A.C. Riddiford, *Electrochemistry and Electrochemical Engineering*, vol. 4 (New York, NY: Interscience, 1966).

57. E. Brunner, *Zeitschrift für Phyzikalische Chemie* 47(1904): p. 56.

58. B. Roald, W. Beck, *Journal of the Electrochemical Society* 98(1951): pp. 277-290.

59. G.I. Taylor, *Proceedings, Royal Society of London* A151(1935): p. 494.

60. A.C. Makrides, N. Hackerman, *Journal of the Electrochemical Society* 105, 3(1958): pp. 156-161.

61. D.J. Robinson, D.R. Gabe, *Transactions of the Institute of Metal Finishing* 48(1970): pp. 35-42.

62. D.R. Gabe, D.J. Robinson, *Transactions of the Institute of Metal Finishing* 49(1971): pp. 17-21.

63. A.J. Arvia, J.S.W. Carrozza, *Electrochimie Acta* 7(1962): pp. 65-78.

64. J.S. Newman, *Electrochemical Systems* (Englewood Cliffs, NJ: Prentice Hall Inc., 1973), pp. 324-325.

65. C.V. King, P.L. Howard, *Industrial and Engineering Chemistry* 29, 1(1937): pp. 75-78.

66. C.V. King, W.H. Cathcart, *Journal of the American Chemical Society* 59(1937): pp. 63-67.

67. R. Kappesser, I. Cornet, R. Grief, *Journal of the Electrochemical Society* 118, 12(1971): pp. 1,957-1,959.

68. P. Harriott, R.M. Hamilton, *Electrochimie Acta* 20(1965): pp. 1,073-1,078.

69. E.S.C. Meyerink, S.K. Friedlander, Chemical Engineering Science 17(1962): p. 121.

70. F.P. Berger, K.F.F.L. Hue, *International Journal of Heat and Mass Transfer* 20(1977): p. 1,185.

71. T.K. Ross, D.H. Jones, *Journal of Applied Chemistry* 12(1962): p. 314.

72. W.L. Friend, A.B. Metzner, *AICHE Journal* 4(1958): p. 393.

73. C.S. Lin, R.W. Moulton, G.L. Putnam, *Industrial and Engineering Chemistry* 45(1953): p. 636.

74. K.G. Diessler, NACA Report 1210 (1955).

75. T.Y. Chen, A. Moccari, D. Macdonald, *Development of Controlled Hydrodynamic Techniques for Corrosion Testing*, MTI Publication T-3, to be published.

76. S.W. Dean, "The Challenge of Predicting Process Corrosion Through Modelling and Laboratory Testing," in *Plant Corrosion*, eds. J.E. Strutt, J.R. Nicholls (Ellis Horwood Publishers, 1987), pp. 19-41.

77. R.D. Armstrong, *Corrosion Science* 11(1971): p. 693.

78. M. Turner, P.A. Brook, *Corrosion Science* 13(1973): p. 973.

79. M.A.C. Decastro, B.E. Wilde, *Corrosion* 35(1979): p. 560.

80. F. Mansfield, J.V. Renkel, *Corrosion* 35(1979): p. 43.

81. F. Franz, P. Novak, *Proceedings of the Conference on Localized Corrosion* (Houston, TX: NACE, 1974), p. 576.

82. N. Sato et al., *Proceedings of the Conference on Localized Corrosion* (Houston, TX: NACE, 1974), p. 447.

83. T. Nakayama et al., *Corrosion* 32(1976): p. 283.

84. F.L. LaQue, *Marine Corrosion* (New York, NY: John Wiley, 1975), p. 138.

85. T.R. Beck, S.G. Chan, *Corrosion* 37(1981): pp. 665-671.

86. D.C. Silverman, "Dimensionless Groups in the Modelling of Corrosion Process." 1984 International Conference on Corrosion Chemistry in Pits, Crevices and Cracks (Teddington, United Kingdom: NPL, 1984).

87. O.T. Fasullo, *Sulphuric Acid Use and Handling* (New York, NY: McGraw-Hill Book Company, 1965), pp. 87-90.

88. J.G. Hines, R.C. Williamson, *Corrosion Science* 4(1964): pp. 201-210.

89. J.G. Hines, R.C. Williamson, *Corrosion Science* 4(1964): pp. 221-235.

90. G. Gilli, F. Zucchi, *Corrosion Science* 8(1968): pp. 801-808.

91. O.T. Fasullo, *Sulphuric Acid Use and Handling* (New York, NY: McGraw-Hill Book Company, 1965), p. 273.

92. G. Schwab, E. Kolb, *Zeitschrift für Physikalische Chemie*, neue folge, 3(1955): p. 52.

93. N.F.H. Bright, H. Hutchinson, D. Smith, *Journal of the Society of the Chemical Industry* 65(1946): pp. 385-388.

94. *International Critical Tables*, vol.5 (New York, NY: McGraw-Hill Book Co., 1929), p. 11.

95. H.L. Silcock, ed., *Solubilities of Inorganic and Organic Compounds*, vol. III, part 1 (Oxford, United Kingdom: Pergammon Press, 1979), pp. 452-466.

96. *Annual Book of ASTM Standards*, 1980 edition, part 12 (Philadelphia, PA: ASTM, 1980), pp. 333-334.

97. L.L. Shrier, ed., *Corrosion*, 1976 edition, vol.1 (London, United Kingdom: Newnes-Butterworth, 1976), pp. 21:63-21:64.

98. L.L. Shrier, ed., *Corrosion*, 1976 edition, vol.1 (London, United Kingdom: Newnes-Butterworth, 1976), pp. 3:86-3:87.

99. E.L. Cussler, *Diffusion, Mass Transfer in Fluid Systems* (Cambridge, United Kingdom: Cambridge University Press, 1984), pp. 288-295.

100. M.G. Fontana, N.D. Greene, *Corrosion Engineering*, 2nd edition (New York, NY: McGraw-Hill Book Co., 1978), p. 225.

101. D.A. Frank-Kamenisti, *Diffusion and Heat Exchange in Chemical Kinetics*, translated from the Russian by N. Thon, (Princeton, NJ: Princeton University Press, 1955), p. 166.

102. I. Cornet, R. Kappesser, *Transaction of the Institute of Chemical Engineers* 47(1969): p. 194.

103. A.J. Ariva, S.W. Carrozza, S.L. Marchiano, *Electrochimie Acta* 9(1964): p. 1,483.

104. T.K. Sherwood, J.M. Ryan, *Chemical Engineering Society* 11(1959): p. 81.

105. T.H. von Karman, *J. Aero. Sc.* 1(1934): p. 1.

106. V.G. Levich, *Physiochemical Hydrodynamics* (Englewood Cliffs, NJ: Prentice Hall Inc., 1962), pp. 144-154.

107. D.C. Silverman, *Corrosion* 40, 5(1984): pp. 220-226.

108. W.M. Kays, *Convective Heat and Mass Transfer* (New York, NY: McGraw-Hill Book Co., 1966).

109. E. Heitz, *Werkstoffe und Korrosion* 15(1964): p. 63.

110. A. Ede, "Effect of 180° Bend on Heat Transfer to Water in a Tube," *Proceedings of the Third International Heat Transfer Conference* 1966, p. 99.

111. E.L. Cussler, *Diffusion, Mass Transfer in Fluid Systems* (Cambridge, United Kingdom: Cambridge University Press, 1984), pp. 282-283.

112. J. Berendson, G. Wranglen, *Corrosion Science* 20(1980): pp. 937-941.

113. *Metals Handbook*, 8th edition, vol. I (Metals Park, OH: ASM, 1961), p. 391.

114. H. Shimada, Y. Sakakibara, H. Okada, *Corrosion* 33(1977): p. 196.

115. E.L. Cussler, *Diffusion, Mass Transfer in Fluid Systems* (Cambridge, United Kingdom: Cambridge University Press, 1984), pp. 185-186.

116. T.K. Ross, A.A. Wragg, *Electrochimie Acta* 10(1965): pp. 1,093-1,106.

117. L. Prandtl, "The Mechanics of Viscous Fluids," *Aerodynamic Theory* vol. 3, W.F. Durand, ed. (Berlin, Germany: Springer, 1935), p. 34.

118. APHA/AWWA/WPCF Standard Methods for the Examination of Waters and Waste Waters, prepared and edited by the American Public Houses Assoc.(APHA), the American Water Works Assoc. (AWWA), and the Water Pollution Control Federation (WPCF), l6th. edition, 1985, pp. 214-219.

APPENDIX A

THEORETICAL DERIVATION OF A RELATION DESCRIBING MASS TRANSFER AT A ROTATING CYLINDER ELECTRODE UNDER TURBULENT FLOW CONDITIONS[13]

Gabe and Robinson have studied mass transfer under turbulent flow conditions at a rotating cylinder electrode (RCE) using the fundamental concepts of hydrodynamics and convective diffusion.[13] They were able to derive theoretically a mass transfer correlation of the type empirically obtained by Eisenberg studying the same system.[46] This appendix is a summary of their theoretical treatment of mass transfer under turbulent flow conditions at an RCE.

Flowing fluids are assumed to be incompressible and Newtonian in nature, so that the shear stress, τ, at any point in a laminar flow is:

$$\tau = \mu \frac{dv}{dy}$$

In turbulent motion, an apparent eddy viscosity, ϵ, is felt,[13] such that:

$$\mu_T = \mu + \rho\epsilon$$

The value of ϵ depends upon the fluid velocity and implies that the friction drag at the liquid/solid interface is greater in turbulent than in laminar flow.[14] The eddying gives rise to enhanced diffusion effects so that the convective diffusion equation for turbulent flow (cf. Section 1.3) may be written as:

$$\bar{v}\nabla c = (D + D_{TURB})\nabla^2 c$$

However, unlike D, D_{TURB} probably varies with flow velocity, i.e., the degree of eddying. Therefore, this equation is only valid with this condition. The Mixing-length theory of Prandtl assumes that $\epsilon = D_{TURB}$ and that turbulence is essentially analogous to the random motion of gas molecules as described by the kinetic theory of gases.[117]

The following treatment is based on the Von Karman theory of the liquid-solid interface that assumes a three-zone model (cf. Figure A.1):[105]

Zone I: a laminar boundary layer where all transfer is due to molecular effects.

Zone II: a viscous transition layer where molecular viscosity tends to damp out the eddies.

Zone III: developed turbulence far from the wall with all transfer due to eddy effects.

Levich has postulated the addition of a fourth layer; namely, the turbulent boundary layer.[106] In this zone, neither molecular viscosity nor diffusion play a noticeable role; both momentum and matter are transferred by means of turbulent eddies. The addition has been advocated to give a smoother transition between Zone III and Zone II.

The three-zone model has found wide acceptance, and Diessler has proposed values for the eddy viscosity for each of the Zones.[74]

In the laminar zone, no eddying occurs:

$$\epsilon = 0$$

In the turbulent main stream flow, ϵ is given by:

$$\epsilon = n^2 yv$$

In the viscous-damping transition layer, the expression suggested for ϵ is:

$$\epsilon = n^2 yv\left[1 - \exp\left(\frac{-n^2 yv}{\nu}\right)\right]$$

The method used by Gabe and Robinson consisted of deriving an expression for D_{TURB}.[13] They then solved Fick's Law for the concentration distribution in each zone and showed that the major resistance to mass transfer occurs in Zone I where molecular effects prevail.

Zone III:

An expression for D_{TURB} given by:

$$D_{TURB} = n^2 yv \arctan\left(a\frac{y}{d}\right)$$

has been derived for Zone III where the turbulent motion is undamped and complete mixing occurs. Thus, there is no concentration gradient and $c_{III} = c_b$.

Zone II:

In Zone II, the expression becomes:

$$D_{TURB} = n^2 y v \arctan\left(a\frac{y}{d}\right)\left[1 - \exp\left(\frac{-n^2 y v}{\nu}\right)\right]$$

that, where y is small and d finite, may be reduced to:

$$D_{TURB} = \frac{b y^3 v^2}{\nu d}$$

where $b = an^4$ is a constant.

If this expression for D_{TURB} is then introduced into the Fick's Law equation, integration between the appropriate limits of Zone II gives:

$$\int_{c_{II}}^{c_b} dc = \int_y^{\delta h} \frac{j}{D_{TURB}} dy$$

and

$$c_{II} = c_b + \frac{j \nu d}{2 b v^2}\left[\frac{1}{\delta h^2} - \frac{1}{y^2}\right] \qquad \text{(A.1)}$$

Zone I:

In Zone I, $y < \delta d$ and $D_{TURB} < D$, molecular diffusion becomes the mechanism of mass transfer so that from Fick's Law:[106]

$$\int_{c_s}^{c_I} dc = \frac{j}{D}\int_o^y dy$$

and

$$c' = c_s + \frac{j}{D}y \tag{A.2}$$

when $y = \delta_d$, $c_I = c_{II}$ and combining equations (A.1) and (A.2) gives:

$$j = \frac{D(c_b - c_s)}{\delta_d - A}$$

where

$$A = \frac{v^2}{4d^2v^2}\,\frac{2d^3}{b}\,\frac{D}{\nu}\left[\frac{1}{\delta_{h^2}} - \frac{1}{\delta_d}\right]$$

the value of A is usually much smaller than δ_d and can be ignored, yielding the simple equation:

$$j = \frac{D(c_b - c_s)}{\delta_d} \tag{A.3}$$

Since $D = D_{TURB}$ when $y = \delta_d$, the equation derived for D_{TURB} in Zone II can give a value for δ_d.[106]

$$\delta_d = \left[\frac{D\nu d}{bv^2}\right]^{\frac{1}{3}}$$

Introducing this into the flux equation (A.3) and converting to current density, it is found that:

$$i = \frac{zFD(c_b - c_s)}{\left[\frac{Dvd}{bv^2}\right]^{\frac{1}{3}}}$$

and after regrouping terms, using dimensionless numbers, it is found that:

$$St = (2b)^{\frac{1}{3}} Re^{-\frac{1}{3}} Sc^{-\frac{2}{3}}$$

Or, using the more widely used Sherwood number (Sh), this is equivalent to:

$$Sh = (2b)^{\frac{1}{3}} Re^{\frac{2}{3}} Sc^{\frac{1}{3}}$$

The equation thus derived is of a similar form to that deduced empirically in several investigations.[46,63,102-104]

The constant $(2b)^{1/3}$ must be derived empirically, and for the concentric cell in which the inner electrode rotates and is under examination, its value is 0.079.[46,61-63]

The treatment used by Gabe and Robinson clearly involves several approximations and assumes a relatively simple model amenable to analytical treatment.[13] The author's own conclusion is, "The analysis yields an equation of the correct form, and any subsequent treatment must be based on a refined model."

In the opinion of the author, the above derivation also highlights (cf. Section 5.1.5) the constant need for experimentation and for the empirical determination of mass transfer correlations applicable to all but the very simple geometries.[44,45]

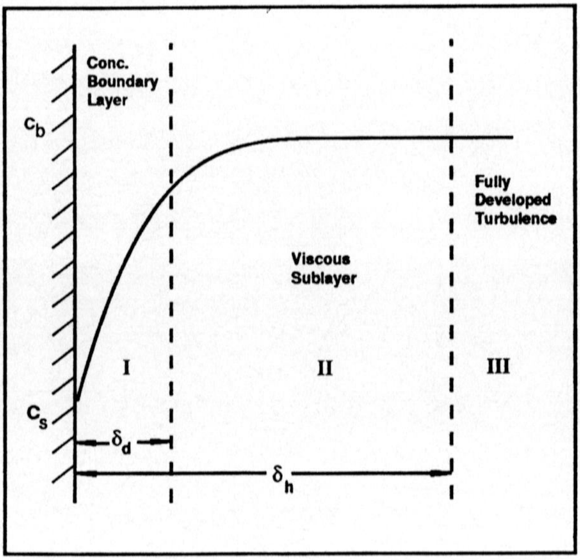

Figure A.1 Model of turbulent flow at an RCE showing a concentration profile over three zones.[19] Reprinted with permission from *Journal of Applied Electrochemistry*, vol. 4, D.R. Gabe, pages 91-108, 1974, Chapman and Hall (London).

APPENDIX B

POLYACRYLAMIDE GEL FOR USE WITH MERCURY/MERCUROUS SULFATE REFERENCE ELECTRODE

SOLUTION NO. 1

50 mL of water + 6 g K_2SO_4

Heat to 35°C (95°F) to dissolve, then allow to cool.

SOLUTION NO. 2

50 mL of water + 15.00 g acrylamide
+ 0.76 g methylene bisacrylamide
+ 0.30 g ammonium persulfate

CATALYST

A stock solution of this can be made up consisting of 1 mL of tetramethylene diamine + 1,000 mL water.

PROCEDURE

Quickly add solution No. 2 to solution No. 1 and deaerate by bubbling with nitrogen for 3 minutes. Then, within 2 to 3 minutes, add 4 mL of the catalyst solution and pour into the probe holder/salt bridge.

APPENDIX C

IRON DETERMINATION BY THE ORTHOPHENANTHROLINE SOLUTION[96]

In this method, the iron is determined photometrically as the orange-red complex that is formed between orthophenanthroline and ferrous ion in a solution buffered at about pH = 4.0. Measurement is made at approximately 490 nm.

(1) Concentration range

The recommended concentration range is from 0.012 mg to 0.24 mg of iron in 100 mL of solution, using a cell depth of 2 cm.

(2) Stability of color

The color is stable for several days.

(3) Interfering elements

Provision is made in the method for the removal of interference due to copper. Copper tends to interfere by forming a complex with the orthophenanthroline, thus preventing complete color development of the iron. However, in the present method sufficient orthophenanthroline is used to permit complete color development in the presence of as much as 2 mg of copper.

(4) Reagents

Use reagents low in iron. Use iron-free distilled water in preparing standards and reagent solutions.

4.1 Hydroxylamine hydrochloride solution (10 g/liter):
Dissolve 1 g of $NH_2OH \cdot HCl$ in water and dilute to 100 mL. Prepare fresh each week as required.

4.2 Freshly prepared iron standard solution (1 mL = 0.02 mg Fe):
Dissolve 0.1405 g of $Fe(NH_4)_2(SO_4)_2 \cdot 6H_2O$ in 100 mL of HCl (1 + 19) and dilute to 1 L in a volumetric flask. Prepare fresh.

4.3 Orthophenanthroline solution (2 g/liter):
Dissolve 0.2 g of orthophenanthroline monohydrate in 90 mL of water with the aid of stirring and gentle heat. Discard the solution if browning occurs. Cool and dilute to 100 mL. This solution is stable for several months.

4.4 Sodium acetate—Acid buffer solution:
Dissolve 270 g of anhydrous sodium acetate in 500 mL water, add 240 mL of acetic acid, cool, and dilute to 1 L.

(5) Preparation of calibration curve

5.1 Calibration solutions:
Transfer 1.0, 2.0, 4.0, 8.0, and 12.0 mL of iron solution (1 mL = 0.02 mg of Fe) to five 100 mL volumetric flasks. Dilute to about 60 mL. Proceed in accordance with 5.3.

5.2 Reference solution:
Add about 60 mL of water to a 100 mL volumetric flask. Proceed in accordance with 5.3.

5.3 Color development:
Add 10 mL of the sodium acetate-acetic acid buffer solution and mix. Add 2 mL of hydroxylamine hydrochloride solution and mix. Add 5 mL of the orthophenanthroline solution, dilute to the mark, and mix.

5.4 Photometry:
Transfer a suitable portion of the reference solution to an absorption cell and adjust the photometer to the initial setting using a light band centered at approximately 490 nm. While maintaining this photometer adjustment, take the photometric readings of the calibration solutions.

5.5 Calibration curve:
Plot the photometric readings of the calibration solutions against milligrams of iron per 100 mL of solution. This is figure C.1.

(6) Procedure

6.1 Reference solution:
Carry a reagent blank through the entire procedure using the same amounts of all the reagents for use as a reference solution.

6.2 Color development:
Transfer the solution, or a suitable aliquot, to a 100 mL volumetric flask. Dilute to 50/60 mL. Add 25 mL of the sodium acetate-acetic acid buffer solution and mix. Add 2 mL of the hydroxylamine hydrochloride solution and mix. Add 10 mL of the orthophenanthroline solution, dilute to the mark, and mix.

6.3 Photometry:
Take the photometric reading of the sample solution as directed in 5.4.

6.4 Calculation:
Convert the photometric readings of the sample or aliquot to milligrams of iron by means of the calibration curve. Calculate the percentage of iron as follows:

$$\text{Iron}\,\% = \frac{A}{(B \cdot 10)}$$

where A = milligrams of iron found in the sample or aliquot used and B = grams of sample, represented by the sample or aliquot used.

(7) Example calculation

According to the calibration curve (Figure C.1), 1 mL of iron solution containing 0.02 mg of iron will give a photometric reading of 0.050.

Using an acid concentration of 98% at 27°C (81°F) (Appendix D, Tables D.6 to D.8 for mild steel), if the spectrophotometer reading is 0.040 (not reported in the Table), the sample will contain:

$$\text{Iron content} = 0.02\,\text{mg} \cdot \frac{0.040}{0.050} = 0.016\,\text{mg Fe}$$

the corresoponding iron % will be:

$$\frac{0.016\,\text{mg}}{1.829\,\text{mg}} \cdot 10 = 0.0008749\%$$

where 1.829 is the acid density, or the number of grams of sample represented by 1 mL of test solution at the concentration and temperature of interest.

The iron molar concentration can then be found as:

$$\frac{8.75 \cdot 10^{-6} \cdot \dfrac{1.829\,\text{g}}{\text{cm}^3}}{\dfrac{55.85\,\text{g}}{\text{mole}}} = \frac{0.287 \cdot 10^{-6}\,\text{moles}}{\text{cm}^3}$$

where 55.85 g/mole is the molecular weight of iron. The value $0.287 \cdot 10^{-6}$ moles/cm^3 is thus reported in Table D.8 (second value) in the c_b

column (bulk acid ferrous ion concentration).

The ferrous ion saturation concentration c_s (also the ferrous sulfate saturation concentration) is calculated from the ferrous sulfate solubility given in percentage form in the appropriate table (cf. also Figure 4.7).[3,95] The calculation is similar to that for the bulk acid ferrous ion concentration.

For 98% acid at 27°C (81°F), the value is read as 0.019%, this can be converted to a molar concentration as follows:

$$0.019\% \cdot \frac{55.85}{151.9} \cdot \frac{1.829}{55.85} = \frac{2.288 \cdot 10^{-6}\,\text{moles}}{\text{cm}^3}$$

where 151.9 g/mole is the molecular weight of $FeSO_4$. At the 68% acid concentration level, the ferrous sulfate film was found to be $FeSO_4 \cdot H_2O$, and the appropriate molecular weight used.[3,95]

(8) Precision and accuracy

Precision and accuracy depend in general on the method of sample collection and storage, the method of color measurement, the iron concentration, and the presence of interfering color, turbidity, and foreign ions. Under optimum conditions, photometric measurement may be reliable to 3% or 3 μg, whichever is greater.[118] Sample variability and instability may limit precision and accuracy of this determination more than will the error of analysis itself. In the present case, no storage of the samples was made as they were immediately treated according to the method described.

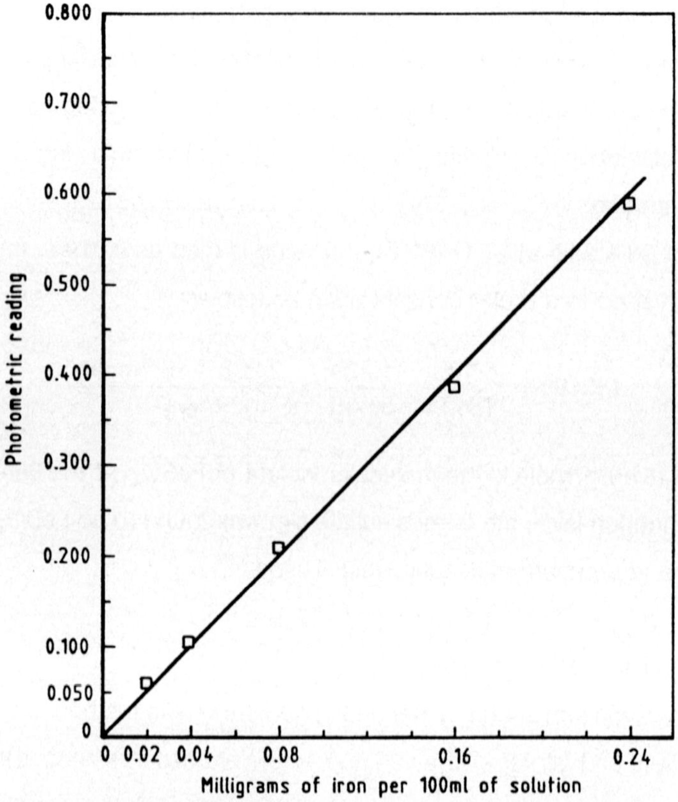

Figure C.1 Calibration curve for the iron determination by the orthophenanthroline method.

APPENDIX D

TABULATED EXPERIMENTAL DATA

Table D.1

Hydrodynamic Corrosion Data for Mild Steel in 68% H_2SO_4 at 27°C (81°F) and 0.81 ms^{-1}

				Alloy MILD STEEL	Solution 68% H_2SO_4	Velocity (ms^{-1}) (rpm) 0.81 500	T° (°K) (°C) 300 27	Exposure Time t (hours) 30	S = 9.74cm² d = 3.1cm
μ : viscosity (kg/ms^{-1}) (CP) 7.6×10^{-3} 7.6		D : diffusion coefficient (m^2/s) 66.3×10^{-12}		ρ : density (kg/m^3) 1,595		C_s surface concentration ($\times 10^{-6}$ mols/cm³) 6.759		F = 96,500 cb M = 55.8 g/mols Z = 2 A = $2.24.10^9 (mols/cm^2/s/mm/y)^{-1}$	
$Re = \frac{dv\rho}{\mu} = \frac{dv}{\nu}$ = 5,282									
$Sc = \frac{\mu}{\rho D} = \frac{\nu}{D}$ = 71,868									
$Sh^{(2)}$: predicted by Eisenberg mass transfer model : 1,706 $Sh^{(2)} = 0.079\ Re^{0.7}\ Sc^{0.356}$				$k^{(2)}$ cm/s = $\frac{ShD}{d}$ = 3.653×10^{-4}		mass transfer factor, $sh^{(2)}/Sc^{0.356}$ = 31.88			
t = time (hours)	I_L measured (mA)	CR (mm/y) measured ($CR^{(1)}$)	Cb ($\times 10^{-6}$ mols/cm³)	$k^{(1)} = \frac{CR^{(1)}}{A(C_s - C_b)}$ cm/s $\times 10^{-4}$	$Sh^{(1)} = \frac{k^{(1)}d}{D}$	I_{LCALC} (mA) $= k^{(2)} z F (C_s - C_b) S$	CR_{CALC} (mm/y) ($CR^{(2)}$)	$\frac{CR^{(1)} - CR^{(2)}}{CR^{(2)}}$(%)	$Sh^{(1)}/Sc^{0.356}$
2	2.20	2.62	2.686	2.870	1,342	2.796	3.327	-21	25.05
8	2.20	2.62	2.795	2.948	1,378	2.720	3.237	-19	25.72
23	2.00	2.38	2.978	2.810	1,313	2.590	3.082	-23	24.51
30	2.00	2.38	3.159	2.951	1,380	2.470	2.940	-19	25.76

$k^{(1)}$, $Sh^{(1)}$: mass transfer factor and Sherwood number determined from limiting current measurement

Table D.2

Hydrodynamic Corrosion Data for Mild Steel in 68% H_2SO_4 at 27°C (81°F) and 2 ms^{-1}

Alloy	Solution	Velocity (ms^{-1}) (rpm)	T° (°K) (°C)	Exposure Time t (hours)	
MILD STEEL	68% H_2SO_4	2 1,235	300 27	46	S = 9.74cm^2 d = 3.1cm

μ : viscosity (kg/ms^{-1}) (CP)	D : diffusion coefficient (m^2/s)	ρ : density (kg/m^3)	C_s surface concentration ($\times 10^{-6}$ $mols/cm^3$)	
7.6×10^{-3} 7.6	66.3×10^{-12}	1,595	6.759	F = 96,500 cb M = 55.8 g/mols Z = 2 A = 2.24.10^9($mols/cm^2/s/mm/y)^{-1}$

$Re = \frac{dv\rho}{\mu} = \frac{dv}{\nu}$ = 13,011

$Sc = \frac{\mu}{\rho D} = \frac{\nu}{D}$ = 71,868

$Sh^{(2)}$: predicted by Eisenberg mass transfer model : 3,210
$Sh^{(2)} = 0.079\ Re^{0.7}\ Sc^{0.356}$

$k^{(2)}$ cm/s = $\frac{ShD}{d}$ = 6.866×10^{-4} mass transfer factor, $sh^{(2)}/Sc^{0.356}$ = 59.92

t=time (hours)	I_L measured (mA)	CR (mm/y) measured ($CR^{(1)}$)	Cb ($\times 10^{-6}$ $mols/cm^3$)	$k^{(1)} = \frac{CR^{(1)}}{A(C_s - C_b)}$ cm/s $\times 10^{-4}$	$Sh^{(1)} = \frac{k^{(1)}d}{D}$	I_{LCALC} (mA) $= k^{(2)}\ z\ F\ (C_s - C_b)\ S$	CR_{CALC} (mm/y) ($CR^{(2)}$)	$\frac{CR^{(1)} - CR^{(2)}}{CR^{(2)}}$(%)	$Sh^{(1)}/Sc^{0.356}$
1	6.70	7.97	0.251	5.47	2,558	8.39	9.98	-20	47.75
4	6.90	8.21	0.358	5.72	2,678	8.26	9.82	-17	49.95
20	6.50	7.73	1.652	6.76	3,161	6.59	7.84	-2	59.00
23	5.50	6.54	1.795	5.88	2,752	6.40	7.62	-14	51.37
30	4.85	5.77	2.010	5.42	2,537	6.13	7.29	-20	47.36
45	3.50	4.17	2.512	4.38	2,047	5.48	6.52	-36	38.21
46	3.50	4.17	2.512	4.38	2,047	5.48	6.52	-36	38.21

$k^{(1)}$, $Sh^{(1)}$: mass transfer factor and Sherwood number determined from limiting current measurement

Table D.3

Hydrodynamic Corrosion Data for Mild Steel in 68% H_2SO_4 at 40°C (104°F) and 0.81 ms^{-1}

Alloy	Solution	Velocity (ms^{-1}) (rpm)	T° (°K) (°C)	Exposure Time t (hours)	
MILD STEEL	68% H_2SO_4	0.812 500	313 40	20	S = 9.74cm^2 d = 3.1cm

μ : viscosity (kg/ms^{-1}) (CP)	D : diffusion coefficient (m^2/s)	ρ : density (kg/m^3)	C_s surface concentration ($\times 10^{-6}$ $mols/cm^3$)	
5.46 x 0.0001 5.46	96.307 x 10^{-12}	1,574	6.67	F = 96,500 cb M = 55.8 g/mols Z = 2 A = 2.24.10^9($mols/cm^2/s/mm/y)^{-1}$

$Re = \frac{dv\rho}{\mu} = \frac{dv}{\nu}$ = 7,256

$Sc = \frac{\mu}{\rho D} = \frac{\nu}{D}$ = 36,018

$Sh^{(2)}$: predicted by Eisenberg mass transfer model : 1,668
$Sh^{(2)} = 0.079\ Re^{0.7}\ Sc^{0.356}$

$k^{(2)}$ cm/s = $\frac{ShD}{d}$ = 5.182 x 10^{-4} mass transfer factor, $sh^{(2)}/Sc^{0.356}$ = 39.82

t = time (hours)	I_L measured (mA)	CR (mm/y) measured ($CR^{[1]}$)	Cb ($\times 10^{-6}$ $mols/cm^3$)	$k^{(1)} = \frac{CR^{(1)}}{A(C_s - C_b)}$ cm/s x 10^{-4}	$Sh^{(1)} = \frac{k^{(1)}d}{D}$	$I_{L_{CALC}}$ (mA) $= k^{(2)}\ z\ F\ (C_s - C_b)\ S$	CR_{CALC} (mm/y) ($CR^{[2]}$)	$\frac{CR^{(1)} - CR^{(2)}}{CR^{(2)}}$(%)	$Sh^{(1)}/Sc^{0.356}$
2	6.75	8.03	1.340	6.72	2,163	5.20	6.18	+29	51.68
4	6.60	7.85	1.650	6.98	2,247	4.88	5.80	+35	53.64
20	4.35	5.17	2.226	5.20	1,672	4.32	5.14	+1	39.96

$k^{(1)}$, $Sh^{(1)}$: mass transfer factor and Sherwood number determined from limiting current measurement

Table D.4

Hydrodynamic Corrosion Data for Mild Steel in 68% H_2SO_4 at 40°C (104°F) and 1.62 ms^{-1}

				Alloy MILD STEEL	Solution 68% H_2SO_4	Velocity (ms^{-1}) (rpm) 1.62 1,000	T° (°K) (°C) 313 40	Exposure Time t (hours) 3	S = 9.74 cm^2 d = 3.1cm
μ : viscosity (kg/ms^{-1}) (CP) 5.46 x 10^{-3} 5.46		D : diffusion coefficient (m^2/s) 96.307 x 10^{-12}		ρ : density (kg/m^3) 1,574		C_s surface concentration (x 10^{-6} $mols/cm^3$) 6.670		F = 96,500 cb M = 55.8 g/mols Z = 2 A = 2.24.10^9($mols/cm^2/s/mm/y$)$^{-1}$	
$Re = \frac{dv\rho}{\mu} = \frac{dv}{\nu}$			= 14,512						
$Sc = \frac{\mu}{\rho D} = \frac{\nu}{D}$			= 36,018						
$Sh^{(2)}$: predicted by Eisenberg mass transfer model : 2,710 $Sh^{(2)} = 0.079\ Re^{0.7}\ Sc^{0.356}$						$k^{(2)}$ cm/s = $\frac{ShD}{d}$ = 8.419 x 10^{-4}	mass transfer factor, $sh^{(2)}/Sc^{0.356}$ = 64.69		
t = time (hours)	I_L measured (mA)	CR (mm/y) measured ($CR^{(1)}$)	Cb (x 10^{-6} $mols/cm^3$)	$k^{(1)} = \frac{CR^{(1)}}{A(C_s - C_b)}$ cm/s x 10^{-4}	$Sh^{(1)} = \frac{k^{(1)}}{D}$	$I_{L_{CALC}}$ (mA) $= k^{(2)}\ z\ F\ (C_s - C_b)\ S$	CR_{CALC} (mm/y) ($CR^{(2)}$)	$\frac{CR^{(1)} - CR^{(2)}}{CR^{(2)}}$(%)	$Sh^{(1)}/Sc^{0.356}$
1	5.6	6.66	2.908	7.90	2,545	5.95	7.08	-6	60.75
3	4.6	5.47	2.944	6.56	2,112	5.89	7.00	-21	51.10
$k^{(1)}$, $Sh^{(1)}$: mass transfer factor and Sherwood number determined from limiting current measurement									

Table D.5

Hydrodynamic Corrosion Data for Mild Steel in 68% H_2SO_4 at 40°C (104°F) and 2 ms^{-1}

				Alloy MILD STEEL	Solution 68% H_2SO_4	Velocity (ms^{-1}) (rpm) 2 1,235	T° (°K) (°C) 313 40	Exposure Time t (hours) 6	S = 9.74cm² d = 3.1cm
μ : viscosity (kg/ms^{-1}) (CP) 5.46×10^{-3} 5.46		D : diffusion coefficient (m^2/s) 96.307×10^{-12}		ρ : density (kg/m^3) 1,574		C_s surface concentration ($\times 10^{-6}$ $mols/cm^3$) 6.670		F = 96,500 cb M = 55.8 g/mols Z = 2 A = $2.24.10^9 (mols/cm^2/s/mm/y)^{-1}$	
$Re = \frac{dv\rho}{\mu} = \frac{dv}{\nu}$		= 17,873							
$Sc = \frac{\mu}{\rho D} = \frac{\nu}{D}$		= 36,018							
$Sh^{(2)}$: predicted by Eisenberg mass transfer model : 3,135 $Sh^{(2)} = 0.079\ Re^{0.7}\ Sc^{0.356}$				$k^{(2)}$ cm/s = $\frac{ShD}{d}$ = 9.74 10^{-4}		mass transfer factor, $sh^{(2)}/Sc^{0.356}$ = 74.84			
t=time (hours)	I_L measured (mA)	CR (mm/y) measured ($CR^{[1]}$)	Cb ($\times 10^{-6}$ $mols/cm^3$)	$k^{(1)} = \frac{CR^{(1)}}{A(C_s - C_b)}$ cm/s $\times 10^{-4}$	$Sh^{(1)} = \frac{k^{(1)}}{D}$	$I_{L_{CALC}}$ (mA) $= \frac{k^{(2)} z F}{(C_s - C_b) S}$	CR_{CALC} (mm/y) ($CR^{[2]}$)	$\frac{CR^{(1)} - CR^{(2)}}{CR^{(2)}}$ (%)	$Sh^{(1)}/Sc^{0.356}$
2	7.7	9.16	2.441	9.67	3,113	7.74	9.21	-1	74.10
3	6.5	7.73	2.657	8.60	2,770	7.34	8.73	-11	66.60
6	6.1	7.25	2.729	8.22	2,646	7.21	8.58	-15	63.61
$k^{(1)}$, $Sh^{(1)}$: mass transfer factor and Sherwood number determined from limiting current measurement									

Table D.6

Hydrodynamic Corrosion Data for Mild Steel in 98% H_2SO_4 at 27°C (81°F) and 0.81 ms^{-1}

	Alloy	Solution	Velocity (ms^{-1}) (rpm)	T° (°K) (°C)	Exposure Time t (hours)	
	MILD STEEL	98% H_2SO_4	0.81 500	300 27	48	S = 9.74 cm^2 d = 3.1 cm

μ : viscosity (kg/ms^{-1}) (CP)	D : diffusion coefficient (m^2/s)	ρ : density (kg/m^3)	C_s surface concentration ($\times 10^{-6}$ $mols/cm^3$)	
19.5×10^{-3} 19.5	25.85×10^{-12}	1,829	2.288	F = 96,500 cb M = 55.8 g/mols Z = 2 A = $2.24 \cdot 10^{9} (mols/cm^2/s/mm/y)^{-1}$

$Re = \frac{dv\rho}{\mu} = \frac{dv}{v}$ = 2,360

$Sc = \frac{\mu}{\rho D} = \frac{v}{D}$ = 412,440

$Sh^{(2)}$: predicted by Eisenberg mass transfer model : 3,118
$Sh^{(2)} = 0.079\ Re^{0.7}\ Sc^{0.356}$

$k^{(2)}$ cm/s = $\frac{ShD}{d}$ = 2.599×10^{-4} mass transfer factor, $sh^{(2)}/Sc^{0.356}$ = 31.25

t = time (hours)	I_L measured (mA)	CR (mm/y) measured ($CR^{(1)}$)	Cb ($\times 10^{-6}$ $mols/cm^3$)	$k^{(1)} = \frac{CR^{(1)}}{A(C_s - C_b)}$ cm/s $\times 10^{-4}$	$Sh^{(1)} = \frac{k^{(1)}d}{D}$	$I_{L_{CALC}}$ (mA) $= k^{(2)}\ z\ F\ (C_s - C_b)\ S$	CR_{CALC} (mm/y) ($CR^{(2)}$)	$\frac{CR^{(1)} - CR^{(2)}}{CR^{(2)}}$ (%)	$Sh^{(1)}/Sc^{0.356}$
1	0.670	0.79	1.075	2.93	3,513	0.593	0.71	+12	35.20
4	0.628	0.74	1.111	2.83	3,393	0.575	0.69	+7	34.00
9	0.620	0.73	1.147	2.88	3,453	0.557	0.66	+10	34.60
24	0.650	0.77	1.219	3.21	3,850	0.522	0.62	+24	38.58
33	0.620	0.73	1.255	3.15	3,777	0.504	0.60	+21	37.85
48	0.610	0.72	1.255	3.11	3,730	0.504	0.60	+20	37.38

$k^{(1)}$, $Sh^{(1)}$: mass transfer factor and Sherwood number determined from limiting current measurement

Table D.7

Hydrodynamic Corrosion Data for Mild Steel in 98% H_2SO_4 at 27°C (81°F) and 1.62 ms^{-1}

	Alloy	Solution	Velocity (ms^{-1}) (rpm)	T° (°K) (°C)	Exposure Time t (hours)	S =9.74cm^2 d =3.1cm
	MILD STEEL	98% H_2SO_4	1.62 1,000	300 27	31	

- μ : viscosity (kg/ms^{-1}) (CP): 19.5 x 10^{-3} 19.5
- D : diffusion coefficient (m^2/s): 25.85 x 10^{-12}
- ρ : density (kg/m^3): 1,829
- C_s surface concentration (x 10^{-6} $mols/cm^3$): 2.288
- F = 96,500 cb M = 55.8 g/mols; Z = 2; A = $2.24.10^9 (mols/cm^2/s/mm/y)^{-1}$

$Re = \frac{dv\rho}{\mu} = \frac{dv}{v}$ = 4,725

$Sc = \frac{\mu}{\rho D} = \frac{v}{D}$ = 412,440

$Sh^{(2)}$: predicted by Eisenberg mass transfer model : 5,321 $Sh^{(2)} = 0.079\ Re^{0.7}\ Sc^{0.356}$

$k^{(2)}$ cm/s = $\frac{ShD}{d}$ = 4.437 x 10^{-4} mass transfer factor, $sh^{(2)}/Sc^{0.356}$ = 53.32

t=time (hours)	I_L measured (mA)	CR (mm/y) measured ($CR^{[1]}$)	Cb (x 10^{-6} $mols/cm^3$)	$k^{(1)} = \frac{CR^{(1)}}{A(C_s - C_b)}$ cm/s x 10^{-4}	$Sh^{(1)} = \frac{k^{(1)}d}{D}$	$I_{L_{CALC}}$ (mA) $= k^{(2)} z F (C_s - C_b) S$	CR_{CALC} (mm/y) ($CR^{[2]}$)	$\frac{CR^{(1)} - CR^{(2)}}{CR^{(2)}}(\%)$	$Sh^{(1)}/Sc^{0.356}$
1	0.945	1.12	1.255	4.84	5,804	0.862	1.03	+9	58.16
8	0.900	1.07	1.309	4.88	5,852	0.816	0.97	+10	58.65
24	0.885	1.05	1.380	5.16	6,188	0.757	0.90	+16	62.01
31	0.855	1.01	1.506	5.76	6,907	0.652	0.78	+29	69.22

$k^{(1)}$, $Sh^{(1)}$: mass transfer factor and Sherwood number determined from limiting current measurement

Table D.8

Hydrodynamic Corrosion Data for Mild Steel in 98% H_2SO_4 at 27°C (81°F) and 2 ms^{-1}

				Alloy MILD STEEL	Solution 98% H_2SO_4	Velocity (ms^{-1}) (rpm) 2 1,235	T° (°K) (°C) 300 27	Exposure Time t (hours)	S = 9.74cm^2 d = 3.1cm
μ : viscosity (kg/ms^{-1}) (CP) 19.5 x 10^{-3} 19.5		D : diffusion coefficient (m^2/s) 25.85 x 10^{-12}		ρ : density (kg/m^3) 1,829		C_s surface concentration (x 10^{-6} $mols/cm^3$) 2.288		F = 96,500 cb M = 55.8 g/mols Z = 2 A = 2.24.10^9($mols/cm^2/s/mm/y$)$^{-1}$	
$Re = \frac{dv\rho}{\mu} = \frac{dv}{\nu}$			= 5,815						
$Sc = \frac{\mu}{\rho D} = \frac{\nu}{D}$			= 412,440						
$Sh^{(2)}$: predicted by Eisenberg mass transfer model : 6,243 $Sh^{(2)} = 0.079\ Re^{0.7}\ Sc^{0.356}$					$k^{(2)}$ cm/s = $\frac{ShD}{d}$ = 5.206 x 10^4	mass transfer factor, $sh^{(2)}/Sc^{0.356}$ = 62.56			
t=time (hours)	I_L measured (mA)	CR (mm/y) measured ($CR^{(1)}$)	Cb (x 10^{-6} $mols/cm^3$)	$k^{(1)} = \frac{CR^{(1)}}{A(C_s - C_b)}$ cm/s x 10^{-4}	$Sh^{(1)} = \frac{k^{(1)}d}{D}$	$I_{L_{CALC}}$ (mA) $= k^{(2)}\ z\ F\ (C_s - C_b)\ S$	CR_{CALC} (mm/y) ($CR^{(2)}$)	$\frac{CR^{(1)} - CR^{(2)}}{CR^{(2)}}$(%)	$Sh^{(1)}/Sc^{0.356}$
7	1.95	2.32	0.144	4.83	5,792	2.10	2.50	-7	58.04
27	1.85	2.20	0.287	4.91	5,888	1.96	2.33	-6	59.00
45	1.55	1.84	0.646	5.00	5,996	1.61	1.91	-4	60.09
51	1.55	1.84	0.717	5.23	6,271	1.54	1.83	0.00	62.84
69	1.45	1.72	0.860	5.37	6,439	1.40	1.67	+3	64.53
76	1.38	1.64	0.896	5.25	6,307	1.37	1.63	+1	63.20
92	1.33	1.58	1.075	5.81	6,967	1.19	1.42	+11	69.91

$k^{(1)}$, $Sh^{(1)}$: mass transfer factor and Sherwood number determined from limiting current measurement

Table D.9

Hydrodynamic Corrosion Data for Mild Steel in 98% H_2SO_4 at 40°C (104°F) and 0.81 ms^{-1}

Alloy	Solution	Velocity (ms^{-1}) (rpm)	T° (°K) (°C)	Exposure Time t (hours)	S = 9.74 cm^2, d = 3.1 cm
MILD STEEL	98% H_2SO_4	0.812 500	313 40	88	

μ : viscosity (kg/ms^{-1}) (CP)	D : diffusion coefficient (m^2/s)	ρ : density (kg/m^3)	C_s surface concentration ($\times 10^{-6}$ $mols/cm^3$)	F = 96,500 cb M = 55.8 g/mols, Z = 2, A = $2.24.10^9 (mols/cm^2/s/mm/y)^{-1}$
13.75×10^{-3} 13.75	38.24×10^{-12}	1,814	4.534	

$Re = \frac{dv\rho}{\mu} = \frac{dv}{\nu}$ = 3,321

$Sc = \frac{\mu}{\rho D} = \frac{\nu}{D}$ = 198,220

$Sh^{(2)}$: predicted by Eisenberg mass transfer model : 4,321
$Sh^{(2)} = 0.079\ Re^{0.7}\ Sc^{0.356}$

$k^{(2)}$ cm/s = $\frac{ShD}{d}$ = 5.33×10^{-4} mass transfer factor, $sh^{(2)}/Sc^{0.356}$ = 56.21

t = time (hours)	I_L measured (mA)	CR (mm/y) measured ($CR^{[1]}$)	Cb ($\times 10^{-6}$ $mols/cm^3$)	$k^{(1)} = \frac{CR^{(1)}}{A(C_s - C_b)}$ cm/s $\times 10^{-4}$	$Sh^{(1)} = \frac{k^{(1)}d}{D}$	I_{LCALC} (mA) $= k^{(2)} z F (C_s - C_b) S$	CR_{CALC} (mm/y) ($CR^{[2]}$)	$\frac{CR^{(1)} - CR^{(2)}}{CR^{(2)}}$(%)	$Sh^{(1)}/Sc^{0.356}$
18	2.61	3.10	2.043	5.55	4,499	2.49	2.96	+5	58.50
45	2.30	2.73	2.545	6.12	4,961	1.99	2.37	+15	64.50
66	2.00	2.38	2.975	6.81	5,520	1.56	1.85	+28	71.80
68	2.75	3.27	1.505	4.82	3,907	3.03	3.60	-9	50.83
73	2.65	3.15	1.613	4.81	3,899	2.92	3.47	-9	50.70
88	2.35	2.79	1.792	4.54	3,680	2.74	3.26	-14	47.87

$k^{(1)}$, $Sh^{(1)}$: mass transfer factor and Sherwood number determined from limiting current measurement

Table D.10

Hydrodynamic Corrosion Data for Mild Steel in 98% H_2SO_4 at 40°C (104°F) and 1.62 ms^{-1}

				Alloy MILD STEEL	Solution 98% H_2SO_4	Velocity (ms^{-1}) (rpm) 1.62 1,000	T° (°K) (°C) 313 40	Exposure Time t (hours) 3	S = 9.74cm^2 d = 3.1cm
μ : viscosity (kg/ms^{-1}) (CP) 13.75 x 10^{-3} 13.75		D : diffusion coefficient (m^2/s) 38.24 x 10^{-12}		ρ : density (kg/m^3) 1,814		C_s surface concentration (x 10^{-6} $mols/cm^3$) 4.534		F = 96,500 cb M = 55.8 g/mols Z = 2 A = 2.24.10^9($mols/cm^2/s/mm/y$)$^{-1}$	
$Re = \frac{dv\rho}{\mu} = \frac{dv}{\nu}$ = 6,642									
$Sc = \frac{\mu}{\rho D} = \frac{\nu}{D}$ = 198,220									
$Sh^{(2)}$: predicted by Eisenberg mass transfer model : 7,576 $Sh^{(2)} = 0.079\ Re^{0.7}\ Sc^{0.356}$						$k^{(2)}$ cm/s = $\frac{ShD}{d}$ = 9.345 x 10^{-4} mass transfer factor, $sh^{(2)}/Sc^{0.356}$ = 98.55			
t=time (hours)	i_L measured (mA)	CR (mm/y) measured ($CR^{[1]}$)	Cb (x 10^{-6} $mols/cm^3$)	$k^{(1)} = \frac{CR^{(1)}}{A(C_s - C_b)}$ cm/s x 10^{-4}	$Sh^{(1)} = \frac{k^{(1)}d}{D}$	$I_{L_{CALC}}$ (mA) $= k^{(2)}\ z\ F\ (C_s - C_b)\ S$	CR_{CALC} (mm/y) ($CR^{[2]}$)	$\frac{CR^{(1)} - CR^{(2)}}{CR^{(2)}}$(%)	$Sh^{(1)}/Sc^{0.356}$
1	4.05	4.82	1.599	7.33	5,942	5.15	6.12	-21	77.29
3	4.35	5.18	1.635	7.96	6,453	5.07	6.03	-14	84.11
$k^{(1)}$, $Sh^{(1)}$: mass transfer factor and Sherwood number determined from limiting current measurement									

Table D.11

Hydrodynamic Corrosion Data for Mild Steel in 98% H_2SO_4 at 40°C (104°F) and 2 ms^{-1}

				Alloy MILD STEEL	Solution 98% H_2SO_4	Velocity (ms^{-1}) (rpm) 2 1,235	T° (°K) (°C) 313 40	Exposure Time t (hours) 53	S = 9.74cm² d = 3.1cm
μ : viscosity (kg/ms^{-1}) (CP) 13.75 x 10^{-3} 13.75		D : diffusion coefficient (m^2/s) 38.24 x 10^{-12}		ρ : density (kg/m^3) 1,814		C_s surface concentration (x 10^{-6} mols/cm³) 4.534		F = 96,500 cb M = 55.8 g/mols Z = 2 A = 2.24.10^9(mols/cm²/s/mm/y)$^{-1}$	
$Re = \frac{dv\rho}{\mu} = \frac{dv}{\nu}$ = 8,180									
$Sc = \frac{\mu}{\rho D} = \frac{\nu}{D}$ = 198,220									
$Sh^{(2)}$: predicted by Eisenberg mass transfer model : 8,968 $Sh^{(2)} = 0.079\,Re^{0.7}\,Sc^{0.356}$				$k^{(2)}$ cm/s = $\frac{ShD}{d}$ = 11.06 x 10^{-4}		mass transfer factor, $sh^{(2)}/Sc^{0.356}$ = 116.66			
t = time (hours)	i_L measured (mA)	CR (mm/y) measured ($CR^{[1]}$)	Cb (x 10^{-6} mols/cm³)	$k^{(1)}$ = $\frac{CR^{(1)}}{A(C_s - C_b)}$ cm/s x 10^{-4}	$Sh^{(1)} = \frac{k^{(1)}d}{D}$	I_{LCALC} (mA) $= k^{(2)}\,z\,F\,(C_s - C_b)\,S$	CR_{CALC} (mm/y) ($CR^{[2]}$)	$\frac{CR^{(1)} - CR^{(2)}}{CR^{(2)}}$(%)	$Sh^{(1)}/Sc^{0.356}$
4	4.59	5.46	1.864	9.03	7,320	5.61	6.67	-18	95.22
23	3.98	4.73	2.508	10.42	8,447	4.21	5.00	-6	109.88
29	3.89	4.63	2.709	11.32	9,176	3.79	4.51	+3	119.37
30	5.22	6.21	1.412	8.88	7,199	6.49	7.72	-19	93.65
36	4.95	5.89	1.612	8.99	7,288	6.07	7.22	-18	94.80
53	4.45	5.29	2.078	9.64	7,815	5.09	6.05	-12	101.66

$k^{(1)}$, $Sh^{(1)}$: mass transfer factor and Sherwood number determined from limiting current measurement

Table D.12

Hydrodynamic Corrosion Data for Mild Steel in 93% H_2SO_4 at 27°C (81°F) and 0.81 ms^{-1}

				Alloy MILD STEEL	Solution 93% H_2SO_4	Velocity (ms^{-1}) (rpm) 0.81 500	T° (°K) (°C) 300 27	Exposure Time t (hours) 43	S = 9.74 cm^2 d = 3.1cm
μ : viscosity (kg/ms^{-1}) (CP) 18.25 x 10^{-3} 18.25		D : diffusion coefficient (m^2/s) 27.616 x 10^{-12}		ρ : density (kg/m^3) 1,825		C_s surface concentration (x 10^{-6} mols/cm^3) 10.816		F = 96,500 cb M = 55.8 g/mols Z = 2 A = 2.24.10^9(mols/cm^2/s/mm/y)$^{-1}$	
$Re = \frac{dv\rho}{\mu} = \frac{dv}{\nu}$ = 2,517									
$Sc = \frac{\mu}{\rho D} = \frac{\nu}{D}$ = 362,009									
$Sh^{(2)}$: predicted by Eisenberg mass transfer model : 3,128 $Sh^{(2)} = 0.079\ Re^{0.7}\ Sc^{0.356}$					$k^{(2)}$ cm/s = $\frac{ShD}{d}$ = 2.786 x 10^{-4} mass transfer factor, $sh^{(2)}/Sc^{0.356}$ = 32.84				
t = time (hours)	i_L measured (mA)	CR (mm/y) measured ($CR^{[1]}$)	Cb (x 10^{-6} mols/cm^3)	$k^{(1)} = \frac{CR^{(1)}}{A(C_s - C_b)}$ cm/s x 10^{-4}	$Sh^{(1)} = \frac{k^{(1)}d}{D}$	$i_{L_{CALC}}$ (mA) $= k^{(2)}\ z\ F\ (C_s - C_b)\ S$	CR_{CALC} (mm/y) ($CR^{[2]}$)	$\frac{CR^{(1)} - CR^{(2)}}{CR^{(2)}}$(%)	$Sh^{(1)}/Sc^{0.356}$
1	4.85	5.77	0.322	2.454	2,755	5.50	6.54	-12	28.92
3	5.00	5.95	0.358	2.540	2,851	5.48	6.52	-9	29.93
18	5.90	7.02	0.774	3.120	3,502	5.26	6.26	+12	36.76
20	5.90	7.02	0.860	3.148	3,533	5.22	6.20	+13	37.08
26	5.40	6.42	1.111	2.953	3,315	5.08	6.05	+6	34.80
43	5.12	6.09	1.541	2.931	3,290	4.86	5.78	+5	34.54

$k^{(1)}$, $Sh^{(1)}$: mass transfer factor and Sherwood number determined from limiting current measurement

Table D.13

Hydrodynamic Corrosion Data for Mild Steel in 93% H_2SO_4 at 27°C (81°F) and 1.62 ms^{-1}

				Alloy MILD STEEL	Solution 93% H_2SO_4	Velocity (ms^{-1}) (rpm) 1.62 1,000	T° (°K) (°C) 300 27	Exposure Time t (hours) 24	S = 9.74 cm^2 d = 3.1cm
μ : viscosity (kg/ms^{-1}) (CP) 18.25 x 10^{-3} 18.25		D : diffusion coefficient (m^2/s) 27.616 x 10^{-12}		ρ : density (kg/m^3) 1,825		C_s surface concentration (x 10^{-6} $mols/cm^3$) 10.816		F = 96,500 cb M = 55.8 g/mols Z = 2 A = 2.24.10^9($mols/cm^2/s/mm/y$)$^{-1}$	
$Re = \frac{dv\rho}{\mu} = \frac{dv}{\nu}$		= 5,034							
$Sc = \frac{\mu}{\rho D} = \frac{\nu}{D}$		= 362,009							
$Sh^{(2)}$: predicted by Eisenberg mass transfer model : 5,333 $Sh^{(2)} = 0.079\ Re^{0.7}\ Sc^{0.356}$				$k^{(2)}$ cm/s = $\frac{ShD}{d}$ = 4.751 x 10^{-4}		mass transfer factor, $sh^{(2)}/Sc^{0.356}$ = 55.98			
t=time (hours)	i_L measured (mA)	CR (mm/y) measured (CR[1])	Cb (x 10^{-6} $mols/cm^3$)	$k^{(1)} = \frac{CR^{(1)}}{A(C_s - C_b)}$ cm/s x 10^{-4}	$Sh^{(1)} = \frac{k^{(1)}d}{D}$	$I_{L_{CALC}}$ (mA) $= k^{(2)} z F (C_s - C_b) S$	CR_{CALC} (mm/y) (CR[2])	$\frac{CR^{(1)} - CR^{(2)}}{CR^{(2)}}$(%)	$Sh^{(1)}/Sc^{0.356}$
1	8.50	10.11	1.541	4.866	5,462	8.28	9.85	+3	57.33
3	8.30	9.87	1.627	4.795	5,382	8.20	9.76	+1	56.49
5	8.30	9.87	1.720	4.844	5,437	8.12	9.67	+2	57.07
8	7.60	9.04	1.870	4.511	5,063	7.99	9.50	-5	53.15
22	6.80	8.09	2.479	4.332	4,862	7.45	8.86	-9	51.04
24	6.80	8.09	2.615	4.403	4,942	7.32	8.71	-7	51.88

$k^{(1)}$, $Sh^{(1)}$: mass transfer factor and Sherwood number determined from limiting current measurement

Table D.14

Hydrodynamic Corrosion Data for Mild Steel in 93% H_2SO_4 at 27°C (81°F) and 2 ms^{-1}

				Alloy MILD STEEL	Solution 93% H_2SO_4	Velocity (ms^{-1}) (rpm) 2 1,235	T° (°K) (°C) 300 27	Exposure Time t (hours) 17	S = 9.74cm^2 d = 3.1cm
μ : viscosity (kg/ms^{-1}) (CP) 18.25 x 10^{-3} 18.25		D : diffusion coefficient (m^2/s) 27.616 x 10^{-12}		ρ : density (kg/m^3) 1,825		C_s surface concentration (x 10^{-6} $mols/cm^3$) 10.816		F = 96,500 cb M = 55.8 g/mols Z = 2 A = 2.24.10^9($mols/cm^2/s/mm/y)^{-1}$	
$Re = \frac{dv\rho}{\mu} = \frac{dv}{\nu}$ = 6,201									
$Sc = \frac{\mu}{\rho D} = \frac{\nu}{D}$ = 362,009									
$Sh^{(2)}$: predicted by Eisenberg mass transfer model : 6,262 $Sh^{(2)} = 0.079\,Re^{0.7}\,Sc^{0.356}$				$k^{(2)}$ cm/s = $\frac{ShD}{d}$ = 5.577 x 10^{-4} mass transfer factor, $sh^{(2)}/Sc^{0.356}$ = 65.74					
t = time (hours)	i_L measured (mA)	CR (mm/y) measured ($CR^{(1)}$)	Cb (x 10^{-6} $mols/cm^3$)	$k^{(1)} = \frac{CR^{(1)}}{A(C_s - C_b)}$ cm/s x 10^{-4}	$Sh^{(1)} = \frac{k^{(1)}d}{D}$	$i_{L_{CALC}}$ (mA) $= k^{(2)}\,z\,F\,(C_s - C_b)\,S$	CR_{CALC} (mm/y) ($CR^{(2)}$)	$\frac{CR^{(1)} - CR^{(2)}}{CR^{(2)}}$(%)	$Sh^{(1)}/Sc^{0.356}$
1	8.50	10.11	1.290	4.740	5,320	9.98	11.88	-15	55.85
3	8.70	10.35	1.333	4.870	5,469	9.94	11.83	-12	57.41
17	7.80	9.28	2.329	4.881	5,479	8.90	10.59	-12	57.51
$k^{(1)}$, $Sh^{(1)}$: mass transfer factor and Sherwood number determined from limiting current measurement									

Table D.15

Hydrodynamic Corrosion Data for Mild Steel in 93% H_2SO_4 at 40°C (104°F) and 0.81 ms^{-1}

		Alloy MILD STEEL	Solution 93% H_2SO_4	Velocity (ms^{-1}) (rpm) 0.81 500	T° (°K) (°C) 313 40	Exposure Time t (hours) 28	S = 9.74cm² d = 3.1cm
μ : viscosity (kg/ms^{-1}) (CP) 12.9×10^{-3} 12.9	D : diffusion coefficient (m^2/s) 40.763×10^{-12}	ρ : density (kg/m^3) 1,814		C_s surface concentration ($\times 10^{-6}$ mols/cm³) 9.553		F = 96,500 cb M = 55.8 g/mols Z = 2 A = $2.24.10^{9}$(mols/cm²/s/mm/y)$^{-1}$	

$Re = \frac{dv\rho}{\mu} = \frac{dv}{v}$ = 3,540

$Sc = \frac{\mu}{\rho D} = \frac{v}{D}$ = 174,456

$Sh^{(2)}$: predicted by Eisenberg mass transfer model : 4,349; $Sh^{(2)} = 0.079\ Re^{0.7}\ Sc^{0.356}$

$k^{(2)}$ cm/s = $\frac{ShD}{d}$ mass transfer factor, $sh^{(2)}/Sc^{0.356}$ = 59.19; = 5.718×10^{-4}

t = time (hours)	I_L measured (mA)	CR (mm/y) measured ($CR^{(1)}$)	Cb ($\times 10^{-6}$ mols/cm³)	$k^{(1)} = \frac{CR^{(1)}}{A(C_s - C_b)}$ cm/s $\times 10^{-4}$	$Sh^{(1)} = \frac{k^{(1)}d}{D}$	I_{CALC} (mA) $= k^{(2)}\ z\ F\ (C_s - C_b)\ S$	CR_{CALC} (mm/y) ($CR^{(2)}$)	$\frac{CR^{(1)} - CR^{(2)}}{CR^{(2)}}$(%)	$Sh^{(1)}/Sc^{0.356}$
2	11.25	13.38	0.101	6.319	4,805	10.15	12.07	+10	65.41
7	11.35	13.50	0.322	6.528	4,964	9.92	11.80	+14	67.57
21	10.35	12.31	1.266	6.631	5,042	8.90	10.59	+16	68.63
24	10.30	12.25	1.395	6.703	5,097	8.76	10.42	+17	69.38
28	9.70	11.54	1.695	6.556	4,986	8.44	10.04	+15	67.87

$k^{(1)}$, $Sh^{(1)}$: mass transfer factor and Sherwood number determined from limiting current measurement

Table D.16

Hydrodynamic Corrosion Data for Mild Steel in 93% H_2SO_4 at 40°C (104°F) and 1.62 ms^{-1}

				Alloy MILD STEEL	Solution 93% H_2SO_4	Velocity (ms^{-1}) (rpm) 1.62 1,000	T° (°K) (°C) 313 40	Exposure Time t (hours) 4	S = 9.74 cm^2 d = 3.1 cm
μ : viscosity (kg/ms^{-1}) (CP) 12.9×10^{-3} 12.9		D : diffusion coefficient (m^2/s) 40.763×10^{-12}		ρ : density (kg/m^3) 1,814		C_s surface concentration ($\times 10^{-6}$ $mols/cm^3$) 9.553		F = 96,500 cb M = 55.8 g/mols Z = 2 A = $2.24.10^9 (mols/cm^2/s/mm/y)^{-1}$	
$Re = \frac{dv\rho}{\mu} = \frac{dv}{\nu}$ = 7,080									
$Sc = \frac{\mu}{\rho D} = \frac{\nu}{D}$ = 174,456									
$Sh^{(2)}$: predicted by Eisenberg mass transfer model : 7,624 $Sh^{(2)} = 0.079\ Re^{0.7}\ Sc^{0.356}$				$k^{(2)}\ cm/s = \frac{ShD}{d}$ mass transfer factor, $sh^{(2)}/Sc^{0.356}$ = 103.78 = 10.025×10^{-4}					
t=time (hours)	i_L measured (mA)	CR (mm/y) measured ($CR^{(1)}$)	Cb ($\times 10^{-6}$ $mols/cm^3$)	$k^{(1)} = \frac{CR^{(1)}}{A(C_s - C_b)}$ cm/s $\times 10^{-4}$	$Sh^{(1)} = \frac{k^{(1)}d}{D}$	I_{LCALC} (mA) $= k^{(2)}\ z\ F\ (C_s - C_b)\ S$	CR_{CALC} (mm/y) ($CR^{(2)}$)	$\frac{CR^{(1)} - CR^{(2)}}{CR^{(2)}}$(%)	$Sh^{(1)}/Sc^{0.356}$
1	14.40	17.13	1.952	10.060	7,650	14.32	17.04	+0.5	104.14
4	15.00	17.85	2.017	10.570	8,038	14.20	16.90	+6	109.42
$k^{(1)}$, $Sh^{(1)}$: mass transfer factor and Sherwood number determined from limiting current measurement									

Table D.17

Hydrodynamic Corrosion Data for Mild Steel in 93% H_2SO_4 at 40°C (104°F) and 2 ms^{-1}

	Alloy	Solution	Velocity	T°	Exposure Time t	
	MILD STEEL	93% H_2SO_4	(ms^{-1}) (rpm) 2 1,235	(°K) (°C) 313 40	(hours) 3	S = 9.74 cm^2 d = 3.1 cm
μ : viscosity (kg/ms^{-1}) (CP) 12.9×10^{-3} 12.9; D : diffusion coefficient (m^2/s) 40.763×10^{-12}	ρ : density (kg/m^3) 1,814		C_s surface concentration ($\times 10^{-6}$ $mols/cm^3$) 9.553		F = 96,500 cb M = 55.8 g/mols Z = 2 A = $2.24.10^9 (mols/cm^2/s/mm/y)^{-1}$	

$Re = \frac{dv\rho}{\mu} = \frac{dv}{\nu}$ = 8,718

$Sc = \frac{\mu}{\rho D} = \frac{\nu}{D}$ = 174,456

$Sh^{(2)}$: predicted by Eisenberg mass transfer model : 9,024 $Sh^{(2)} = 0.079\ Re^{0.7}\ Sc^{0.356}$

$k^{(2)}$ cm/s = $\frac{ShD}{d}$ = 11.866×10^{-4}; mass transfer factor, $sh^{(2)}/Sc^{0.356}$ = 122.85

t=time (hours)	I_L measured (mA)	CR (mm/y) measured ($CR^{(1)}$)	Cb ($\times 10^{-6}$ $mols/cm^3$)	$k^{(1)}$ = $\frac{CR^{(1)}}{A(C_s - C_b)}$ cm/s $\times 10^{-4}$	$Sh^{(1)} = \frac{k^{(1)}d}{D}$	$I_{L_{CALC}}$ (mA) $= k^{(2)}\, z\, F\, (C_s - C_b)\, S$	CR_{CALC} (mm/y) ($CR^{(2)}$)	$\frac{CR^{(1)} - CR^{(2)}}{CR^{(2)}}$(%)	$Sh^{(1)}/Sc^{0.356}$
1	15.70	18.68	2.396	11.65	8,860	15.96	18.99	-2	120.61
3	16.50	19.63	2.610	12.62	9,597	15.48	18.42	+6	130.64

$k^{(1)}$, $Sh^{(1)}$: mass transfer factor and Sherwood number determined from limiting current measurement

Table D.18

Hydrodynamic Corrosion Data for Mondi A in 98% H_2SO_4 at 27°C (81°F) and 0.81 ms^{-1}

	Alloy	Solution	Velocity (ms^{-1}) (rpm)	T° (°K) (°C)	Exposure Time t (hours)	
	MONDI A	98%H_2SO_4	0.81 500	300 27	24	S =9.74cm^2 d =3.1cm

μ : viscosity (kg/ms^{-1}) (CP)	D : diffusion coefficient (m^2/s)	ρ : density (kg/m^3)	C_s surface concentration ($\times 10^{-6}$ $mols/cm^3$)	
19.5×10^{-3} 19.5	25.85×10^{-12}	1,829	2.288	F = 96,500 cb M = 55.8 g/mols Z = 2 A = $2.24.10^9 (mols/cm^2/s/mm/y)^{-1}$

$Re = \frac{dv\rho}{\mu} = \frac{dv}{\nu}$ = 2,360

$Sc = \frac{\mu}{\rho D} = \frac{\nu}{D}$ = 412,440

$Sh^{(2)}$: predicted by Eisenberg mass transfer model : 1,810
$Sh^{(2)} = 0.079\ Re^{0.7}\ Sc^{0.356}$

$k^{(2)}$ cm/s = $\frac{ShD}{d}$ = 1.51×10^{-4} mass transfer factor, $sh^{(2)}/Sc^{0.356}$ = 18.14

t=time (hours)	I_L measured (mA)	CR (mm/y) measured ($CR^{(1)}$)	Cb ($\times 10^{-6}$ $mols/cm^3$)	$k^{(1)} = \frac{CR^{(1)}}{A(C_s - C_b)}$ cm/s $\times 10^{-4}$	$Sh^{(1)} = \frac{k^{(1)}d}{D}$	I_{LCALC} (mA) $= k^{(2)} z F (C_s - C_b) S$	CR_{CALC} (mm/y) ($CR^{(2)}$)	$\frac{CR^{(1)} - CR^{(2)}}{CR^{(2)}}$(%)	$Sh^{(1)}/Sc^{0.356}$
6	0.435	0.51	0.758	1.508	1,808	0.434	0.516	0	18.12
11	0.380	0.45	0.823	1.377	1,651	0.415	0.494	-8	16.54
24	0.460	0.54	0.895	1.746	2,094	0.395	0.470	+16	20.99

$k^{(1)}$, $Sh^{(1)}$: mass transfer factor and Sherwood number determined from limiting current measurement

Table D.19

Hydrodynamic Corrosion Data for Mondi A in 98% H_2SO_4 at 27°C (81°F) and 1.62 ms^{-1}

	Alloy	Solution	Velocity	T°	Exposure Time t	
	MONDI A	98%H_2SO_4	(ms^{-1}) (rpm) 1.62 1,000	(°K) (°C) 300 27	(hours) 98	S =9.74cm^2 d =3.1cm

μ : viscosity (kg/ms^{-1}) (CP)	D : diffusion coefficient (m^2/s)	ρ : density (kg/m^3)	C_s surface concentration (x 10^{-6} $mols/cm^3$)	
19.5 x 10^{-3} 19.5	25.85 x 10^{-12}	1,829	2.288	F = 96,500 cb M = 55.8 g/mole Z = 2 A = 2.24.10^9($mole/cm^2/s/mm/y)^{-1}$

$Re = \frac{dv\rho}{\mu} = \frac{dv}{\nu}$ = 4,725

$Sc = \frac{\mu}{\rho D} = \frac{\nu}{D}$ = 412,440

$Sh^{(2)}$: predicted by Eisenberg mass transfer model : 2,942
$Sh^{(2)} = 0.079\ Re^{0.7}\ Sc^{0.356}$

$k^{(2)}$ cm/s $= \frac{ShD}{d}$ = 2.453 x 10^{-4} mass transfer factor, $sh^{(2)}/Sc^{0.356}$ = 29.49

t=time (hours)	I_L measured (mA)	CR (mm/y) measured ($CR^{(1)}$)	Cb (x 10^{-6} $mols/cm^3$)	$k^{(1)} = \frac{CR^{(1)}}{A(C_s - C_b)}$ cm/s x 10^{-4}	$Sh^{(1)} = \frac{k^{(1)}d}{D}$	I_{calc} (mA) $= k^{(2)}\ z\ F\ (C_s - C_b)\ S$	CR_{CALC} (mm/y) ($CR^{(2)}$)	$\frac{CR^{(1)} - CR^{(2)}}{CR^{(2)}}$(%)	$Sh^{(1)}/Sc^{0.356}$
5	1.70	2.02	0.213	4.354	5,221	0.956	1.14	+78	52.32
24	0.960	1.14	0.298	2.564	3,074	0.916	1.09	+5	30.80
30	0.820	0.97	0.352	2.250	2,698	0.891	1.06	-8	27.02
37	0.820	0.97	0.429	2.345	2,811	0.853	1.015	-4	28.17
50	0.820	0.97	0.481	2.410	2,891	0.830	0.987	-1	28.97
61	0.750	0.89	0.501	2.228	2,666	0.821	0.977	-9	26.71
74	0.700	0.83	0.615	2.223	2,665	0.769	0.915	-9	26.70
96	0.670	0.79	0.715	2.266	2,717	0.722	0.859	-7	27.22
98	0.670	0.79	0.715	2.266	2,717	0.722	0.859	-7	27.22

$k^{(1)}$, $Sh^{(1)}$: mass transfer factor and Sherwood number determined from limiting current measurement

Table D.20

Hydrodynamic Corrosion Data for Mondi A in 98% H_2SO_4 at 27°C (81°F) and 2 ms^{-1}

		Alloy	Solution	Velocity	T°	Exposure Time t	
		MONDI A	98%H_2SO_4	(ms^{-1}) (rpm) 2 1,235	(°K) (°C) 300 27	(hours) 75	S = 9.74cm^2 d = 3.1cm
μ : viscosity (kg/ms^{-1}) (CP) 19.5 x 10^{-3} 19.5	D : diffusion coefficient (m^2/s) 25.85 x 10^{-12}	ρ : density (kg/m^3) 1,829		C_s surface concentration (x 10^{-6} $mols/cm^3$) 2.288		F = 96,500 cb M = 55.8 g/mols Z = 2 A = 2.24.10^9($mols/cm^2/s/mm/y$)$^{-1}$	

$Re = \frac{dv\rho}{\mu} = \frac{dv}{\nu}$ = 5,815

$Sc = \frac{\mu}{\rho D} = \frac{\nu}{D}$ = 412,440

$Sh^{(2)}$: predicted by Eisenberg mass transfer model : 3,403 $Sh^{(2)} = 0.079\ Re^{0.7}\ Sc^{0.356}$ $k^{(2)}$ cm/s = $\frac{ShD}{d}$ = 2.838 x 10^{-4} mass transfer factor, $sh^{(2)}/Sc^{0.356}$ = 34.10

t = time (hours)	I_L measured (mA)	CR (mm/y) measured ($CR^{(1)}$)	Cb (x 10^{-6} $mols/cm^3$)	$k^{(1)} = \frac{CR^{(1)}}{A(C_s - C_b)}$ cm/s x 10^{-4}	$Sh^{(1)} = \frac{k^{(1)}d}{D}$	I_{CALC} (mA) = $k^{(2)}$ z F (C_s - C_b) S	CR_{CALC} (mm/y) ($CR^{(2)}$)	$\frac{CR^{(1)} - CR^{(2)}}{CR^{(2)}}$(%)	$Sh^{(1)}/Sc^{0.356}$
5	0.660	0.78	0.895	2.517	3,018	0.743	0.884	-11	30.24
23	0.555	0.66	1.005	2.194	2,631	0.684	0.814	-19	26.36
24	0.550	0.65	1.005	2.277	2,731	0.684	0.814	-19.5	27.37
29	0.550	0.65	1.005	2.277	2,731	0.684	0.814	-19.5	27.37
71	0.490	0.58	1.038	2.081	2,496	0.666	0.792	-16.5	25.01
75	0.50 W/e 1	0.59	1.076	2.191	2,629	0.645	0.768	-22	26.35
	0.43 W/e 2	0.51		1.844	2,259			-33	22.63

$k^{(1)}$, $Sh^{(1)}$: mass transfer factor and Sherwood number determined from limiting current measurement

W/e 1: Working electrode 1
W/e 2: Working electrode 2

Table D.21

Hydrodynamic Corrosion Data for Ductile Iron in 98% H_2SO_4 at 27°C (81°F) and 0.81 ms^{-1}

				Alloy DUCTILE IRON	Solution 98%H_2SO_4	Velocity (ms^{-1}) (rpm) 0.81 500	T° (°K) (°C) 300 27	Exposure Time t (hours) 24	S =9.74cm^2 d =3.1cm
μ : viscosity (kg/ms^{-1}) (CP) 19.5 x 10^{-3} 19.5		D : diffusion coefficient (m^2/s) 25.85 x 10^{-12}		ρ : density (kg/m^3) 1,829		C_s surface concentration (x 10^{-6} $mols/cm^3$) 2.288		F = 96,500 cb M = 55.8 g/mols Z = 2 A = 2.24.10^9($mols/cm^2/s/mm/y)^{-1}$	
$Re = \frac{dv\rho}{\mu} = \frac{dv}{\nu}$		= 2,360							
$Sc = \frac{\mu}{\rho D} = \frac{\nu}{D}$		= 412,440							
$Sh^{(2)}$: predicted by Eisenberg mass transfer model : 1,810 $Sh^{(2)} = 0.079\ Re^{0.7}\ Sc^{0.356}$					$K^{(2)}$ cm/s = $\frac{ShD}{d}$ = 1.51 x 10^{-4}	mass transfer factor, $sh^{(2)}/Sc^{0.356}$ = 18.14			
t=time (hours)	I_L measured (mA)	CR (mm/y) measured ($CR^{(1)}$)	Cb (x 10^{-6} $mols/cm^3$)	$k^{(1)} = \frac{CR^{(1)}}{A(C_s - C_b)}$ cm/s x 10^{-4}	$Sh^{(1)} = \frac{k^{(1)}d}{D}$	I_{LCALC} (mA) $= k^{(2)}\ z\ F\ (C_s - C_b)\ S$	CR_{CALC} (mm/y) ($CR^{(2)}$)	$\frac{CR^{(1)} - CR^{(2)}}{CR^{(2)}}$(%)	$Sh^{(1)}/Sc^{0.356}$
4	0.45	0.53	0.644	1.454	1,743	0.467	0.55	-4	17.46
19	0.41	0.48	0.680	1.354	1,623	0.457	0.54	-10	16.26
24	0.40	0.47	0.715	1.350	1,618	0.447	0.53	-10	16.21

$k^{(1)}$, $Sh^{(1)}$: mass transfer factor and Sherwood number determined from limiting current measurement

Table D.22

Hydrodynamic Corrosion Data for Ductile Iron in 98% H_2SO_4 at 27°C (81°F) and 1.62 ms^{-1}

	Alloy	Solution	Velocity	T°	Exposure Time t	
	DUCTILE IRON	98% H_2SO_4	(ms^{-1}) (rpm) 1.62 1,000	(°K) (°C) 300 27	(hours) 28	S = 9.74 cm^2 d = 3.1 cm

μ : viscosity	D : diffusion coefficient	ρ : density (kg/m^3)	C_s surface concentration	
(kg/ms^{-1}) (CP) 19.5 x 10^{-3} 19.5	(m^2/s) 25.85 x 10^{-12}	1,829	(x 10^{-6} $mols/cm^3$) 2.288	F = 96,500 cb M = 55.8 g/mols Z = 2 A = 2.24.10^9($mols/cm^2/s/mm/y$)$^{-1}$

$Re = \frac{dv\rho}{\mu} = \frac{dv}{\nu}$ = 4,725

$Sc = \frac{\mu}{\rho D} = \frac{\nu}{D}$ = 412,440

$Sh^{(2)}$: predicted by Eisenberg mass transfer model : 2,942
$Sh^{(2)} = 0.079\ Re^{0.7}\ Sc^{0.356}$

$K^{(2)}$ cm/s = $\frac{ShD}{d}$ = 2.453 x 10^{-4} mass transfer factor, $sh^{(2)}/Sc^{0.356}$ = 29.49

t = time (hours)	I_L measured (mA)	CR (mm/y) measured ($CR^{(1)}$)	Cb (x 10^{-6} $mols/cm^3$)	$k^{(1)} = \frac{CR^{(1)}}{A(C_s - C_b)}$ cm/s x 10^{-4}	$Sh^{(1)} = \frac{k^{(1)}d}{D}$	I_{CALC} (mA) $= k^{(2)}\ z\ F\ (C_s - C_b)\ S$	CR_{CALC} (mm/y) ($CR^{(2)}$)	$\frac{CR^{(1)} - CR^{(2)}}{CR^{(2)}}$(%)	$Sh^{(1)}/Sc^{0.356}$
7	0.67 w/e 1	0.79	0.574	2.076	2,489	0.790	0.94	-15	24.94
	0.55 w/e 2	0.65		1.704	2,043			-30	20.47
19	0.58 w/e 1	0.69	0.609	1.836	2,202	0.773	0.91	-25	22.06
	0.54 w/e 2	0.64		1.709	2,049			-30	20.53
28	0.55 w/e 1	0.65	0.644	1.777	2,131	0.758	0.90	-28	21.36
	0.53 w/e 2	0.63		1.712	2,053			-30	20.57

$k^{(1)}$, $Sh^{(1)}$: mass transfer factor and Sherwood number determined from limiting current measurement

w/e 1: Working electrode 1
w/e 2: Working electrode 2

Table D.23

Hydrodynamic Corrosion Data for Ductile Iron in 98% H_2SO_4 at 27°C (81°F) and 2 ms^{-1}

	Alloy	Solution	Velocity	T°	Exposure Time t (hours)	
	DUCTILE IRON	98%H_2SO_4	(ms^{-1}) (rpm) 2 1,235	(°K) (°C) 300 27	51	S = 9.74cm^2 d = 3.1cm

μ : viscosity (kg/ms^{-1}) (CP)	D : diffusion coefficient (m^2/s)	ρ : density (kg/m^3)	C_s surface concentration (x 10^{-6} $mols/cm^3$)	
19.5 x 10^{-3} 19.5	25.85 x 10^{-12}	1,829	2.288	F=96,500 cb M = 55.8 g/mols Z=2 A=2.24.10^9($mols/cm^2/s/mm/y)^{-1}$

$$Re = \frac{dv\rho}{\mu} = \frac{dv}{\nu} = 5{,}815$$

$$Sc = \frac{\mu}{\rho D} = \frac{\nu}{D} = 412{,}440$$

$Sh^{(2)}$: predicted by Eisenberg mass transfer model : 3,403
$Sh^{(2)} = 0.079\ Re^{0.7}\ Sc^{0.356}$

$k^{(2)}$ cm/s = $\frac{ShD}{d}$ mass transfer factor, $sh^{(2)}/Sc^{0.356}$ = 34.10
= 2.838 x 10^{-4}

t=time (hours)	I_L measured (mA)	CR (mm/y) measured ($CR^{(1)}$)	Cb (x 10^{-6} $mols/cm^3$)	$k^{(1)} = \frac{CR^{(1)}}{A(C_s - C_b)}$ cm/s x 10^{-4}	$Sh^{(1)} = \frac{k^{(1)}d}{D}$	$I_{L_{CALC}}$ (mA) $= k^{(2)}\ z\ F\ (C_s - C_b)\ S$	CR_{CALC} (mm/y) ($CR^{(2)}$)	$\frac{CR^{(1)} - CR^{(2)}}{CR^{(2)}}$(%)	$Sh^{(1)}/Sc^{0.356}$
1	3.17 w/e 1	3.77							
2	3.12 w/e 2	3.71							
4	2.12 w/e 1	2.52							
10	1.27 w/e 1&2	1.51	0.359	3.497	4,194	1.03	1.22	-23	42.03
22	0.65 w/e 1	0.77	0.430	1.858	2,228	0.99	1.17	-34	22.32
	0.97 w/e 2	1.15		2.773	3,325			-2	33.32
28	0.63 w/e 1	0.74	0.430	1.801	2,159	0.99	1.17	-36	21.63
	0.86 w/e 2	1.02		2.458	2,947			-13	29.53
34	0.60 w/e 1	0.71	0.502	1.784	2,139	0.95	1.13	-37	21.43
	0.77 w/e 2	0.91		2.290	2,746			-19	27.52
45	0.60 w/e 1	0.71	0.574	1.859	2,229	0.91	1.08	-34	22.33
	0.77 w/e 2	0.91		2.386	2,861			-15	28.67
51	0.62 w/e 1	0.73							
	0.84 w/e 2	0.99							

$k^{(1)}$, $Sh^{(1)}$: mass transfer factor and Sherwood number determined from limiting current measurement

w/e 1: Working electrode 1
w/e 2: Working electrode 2

Table D.24

Hydrodynamic Corrosion Data for Grey Cast Iron in 98% H_2SO_4 at 27°C (81°F) and 0.81 ms^{-1}

Alloy	Solution	Velocity (ms^{-1}) (rpm)	T° (°K) (°C)	Exposure Time t (hours)	
GREY CAST IRON	98% H_2SO_4	0.81 500	300 27	31	S = 9.74 cm^2 d = 3.1 cm

μ : viscosity (kg/ms^{-1}) 19.5 x 10^{-3}; (CP) 19.5

D : diffusion coefficient (m^2/s) 25.85 x 10^{-12}

ρ : density (kg/m^3) 1,829

C_s surface concentration (x 10^{-6} $mols/cm^3$) 2.288

F = 96,500 cb M = 55.8 g/mols
Z = 2
A = 2.24.10^9 $(mols/cm^2/s/mm/y)^{-1}$

$$Re = \frac{dv\rho}{\mu} = \frac{dv}{\nu} = 2{,}360$$

$$Sc = \frac{\mu}{\rho D} = \frac{\nu}{D} = 412{,}440$$

$Sh^{(2)}$: predicted by Eisenberg mass transfer model : 1,810
$Sh^{(2)} = 0.079\ Re^{0.7}\ Sc^{0.356}$

$k^{(2)}$ cm/s = $\frac{ShD}{d}$ = 1.51 x 10^{-4}

mass transfer factor, $sh^{(2)}/Sc^{0.356}$ = 18.14

t = time (hours)	I_L measured (mA)	CR (mm/y) measured ($CR^{(1)}$)	Cb (x 10^{-6} $mols/cm^3$)	$k^{(1)} = \frac{CR^{(1)}}{A(C_s - C_b)}$ cm/s x 10^{-4}	$Sh^{(1)} = \frac{k^{(1)}d}{D}$	I_{CALC} (mA) $= k^{(2)} z F (C_s - C_b) S$	CR_{CALC} (mm/y) ($CR^{(2)}$)	$\frac{CR^{(1)} - CR^{(2)}}{CR^{(2)}}$ (%)	$Sh^{(1)}/Sc^{0.356}$
3	0.35 W/e 1	0.41	0.431	1.000	1,199	0.527	0.62	-34	12.02
4	0.41 W/e 2	0.48							
5	0.333 W/e 1	0.39	0.502	0.990	1,187	0.507	0.60	-34	11.89
25	0.330 W/e 1	0.39	0.538	1.000	1,199	0.497	0.59	-34	12.01
	0.40 W/e 2	0.47		1.214	1,456			-20	14.59
31	0.40 W/e 2	0.47	0.610	1.266	1,518	0.476	0.56	-16	15.21

$k^{(1)}$, $Sh^{(1)}$: mass transfer factor and Sherwood number determined from limiting current measurement

W/e 1: Working electrode 1
W/e 2: Working electrode 2

Table D.25

Hydrodynamic Corrosion Data for Grey Cast Iron in 98% H_2SO_4 at 27°C (81°F) and 1.62 ms^{-1}

t=time (hours)	I_L measured (mA)	CR (mm/y) measured ($CR^{(1)}$)	Cb (x 10^{-6} mols/cm³)	$k^{(1)} = \frac{CR^{(1)}}{A(C_s - C_b)}$ cm/s x 10^{-4}	$Sh^{(1)} = \frac{k^{(1)}d}{D}$	I_{CALC} (mA) $= k^{(2)} z F (C_s - C_b) S$	CR_{CALC} (mm/y) ($CR^{(2)}$)	$\frac{CR^{(1)} - CR^{(2)}}{CR^{(2)}}$(%)	$Sh^{(1)}/Sc^{0.356}$
				Alloy GREY CAST IRON	Solution 98% H_2SO_4	Velocity (ms^{-1}) (rpm) 1.62 1,000	T° (°K) (°C) 300 27	Exposure Time t (hours) 25	S = 9.74cm² d = 3.1cm
μ : viscosity (kg/ms⁻¹) (CP) 19.5 x 10^{-3} 19.5		D : diffusion coefficient (m²/s) 25.85 x 10^{-12}		ρ : density (kg/m³) 1,829		C_s surface concentration (x 10^{-6} mols/cm³) 2.288		F=96,500 cb M = 55.8 g/mols Z=2 A=2.24.10⁹(mols/cm²/s/mm/y)⁻¹	
$Re = \frac{dv\rho}{\mu} = \frac{dv}{\nu} = 4{,}725$									
$Sc = \frac{\mu}{\rho D} = \frac{\nu}{D} = 412{,}440$									
$Sh^{(2)}$: predicted by Eisenberg mass transfer model : 2,942 $Sh^{(2)} = 0.079\ Re^{0.7}\ Sc^{0.356}$				$k^{(2)}$ cm/s = $\frac{ShD}{d}$ = 2.453 x 10^{-4}		mass transfer factor, $sh^{(2)}/Sc^{0.356}$ = 29.49			
2	0.425 w/e1	0.50	0.359	1.193	1,431	0.889	1.05	-52	14.34
3	0.590 w/e2	0.70	0.395	1.655	1,985	0.873	1.03	-32	19.89
4	0.430 w/e1	0.51	0.395	1.206	1,446	0.873	1.03	-51	14.49
6	0.600 w/e2	0.71	0.395	1.683	2,019	0.873	1.03	-31	20.23
8	0.450 w/e1	0.53	0.395	1.261	1,513	0.873	1.03	-48	15.16
23	0.600 w/e2	0.71	0.420	1.706	2,045	0.861	1.02	-30	20.49
25	0.420 w/e1	0.49	0.431	1.199	1,439	0.856	1.01	-51	14.42
$k^{(1)}$, $Sh^{(1)}$: mass transfer factor and Sherwood number determined from limiting current measurement									

w/e 1: Working electrode 1
w/e 2: Working electrode 2

Table D.26

Hydrodynamic Corrosion Data for Grey Cast Iron in 98% H_2SO_4 at 27°C (81°F) and 2 ms^{-1}

	Alloy	Solution	Velocity	T°	Exposure Time t	
	GREY CAST IRON	98%H_2SO_4	(ms^{-1}) (rpm) 2 1,235	(°K) (°C) 300 27	(hours) 45	S = 9.74cm^2 d = 3.1cm

μ : viscosity (kg/ms^{-1}) (CP)	D : diffusion coefficient (m^2/s)	ρ : density (kg/m^3)	C_s surface concentration (x 10^{-6} mols/cm^3)	
19.5 x 10^{-3} 19.5	25.85 x 10^{-12}	1,829	2.288	F=96,500 cb M = 55.8 g/mols Z=2 A=2.24.10^9(mols/cm^2/s/mm/y)$^{-1}$

$Re = \frac{dv\rho}{\mu} = \frac{dv}{\nu}$ = 5,815

$Sc = \frac{\mu}{\rho D} = \frac{\nu}{D}$ = 412,440

$Sh^{(2)}$: predicted by Eisenberg mass transfer model : 3,403
$Sh^{(2)} = 0.079\, Re^{0.7}\, Sc^{0.356}$

$k^{(2)}$ cm/s = $\frac{ShD}{d}$ = 2.838 x 10^{-4} mass transfer factor, $sh^{(2)}/Sc^{0.356}$ = 34.10

t=time (hours)	I_L measured (mA)	CR (mm/y) measured ($CR^{(1)}$)	Cb (x 10^{-6} mols/cm^3)	$k^{(1)} = \frac{CR^{(1)}}{A(C_s - C_b)}$ cm/s x 10^{-4}	$Sh^{(1)} = \frac{k^{(1)}}{D}$	I_{CALC} (mA) $= k^{(2)}\, z\, F\, (C_s - C_b)\, S$	CR_{CALC} (mm/y) ($CR^{(2)}$)	$\frac{CR^{(1)} - CR^{(2)}}{CR^{(2)}}$(%)	$Sh^{(1)}/Sc^{0.356}$
2	2.1 W/e 1	2.49	0.100	5.098	6,113	1.16	1.38	+81	61.26
7	1.82 W/e 2	2.16	0.158	4.539	5,443	1.13	1.34	+60	54.55
8	1.58 W/e 1	1.88	0.158	3.940	4,725	1.13	1.34	+40	47.35
22	0.888 W/e1	0.74	0.237	1.618	1,940	1.09	1.29	-42	19.44
	0.625 W/e2	1.05		2.300	2,758			-18	27.64
24	0.580 W/e1	0.69	0.237	1.502	1,801	1.09	1.29	-47	18.05
30	0.870 W/e2	1.03	0.287	2.309	2,769	1.06	1.26	-18	27.75
32	0.530 W/e1	0.63	0.287	1.407	1,687	1.06	1.26	-50	16.90
44	0.700 W/e2	0.83	0.359	1.923	2,306	1.03	1.22	-32	23.11
45	0.430 W/e1	0.51	0.359	1.181	1,416	1.03	1.22	-58	14.19

$k^{(1)}$, $Sh^{(1)}$: mass transfer factor and Sherwood number determined from limiting current measurement

W/e 1: Working electrode 1
W/e 2: Working electrode 2

Table D.27

Hydrodynamic Corrosion Data for Mondi B in 98% H_2SO_4 at 27°C (81°F) and 0.81 ms^{-1}

	Alloy	Solution	Velocity	T°	Exposure Time t	
	MONDI B	98%H_2SO_4	(ms^{-1}) (rpm) 0.81 500	(°K) (°C) 300 27	(hours) 88	S = 9.74cm^2 d = 3.1cm
μ : viscosity (kg/ms^{-1}) (CP) 19.5 x 10^{-3} 19.5	D : diffusion coefficient (m^2/s) 25.85 x 10^{-12}	ρ : density (kg/m^3) 1,829	C_s surface concentration (x 10^{-6} mols/cm^3) 2.288		F=96,500 cb M = 55.8 g/mols Z=2 A=2.24.10^9(mols/cm^2/s/mm/y)$^{-1}$	

$Re = \frac{dv\rho}{\mu} = \frac{dv}{\nu}$ = 2,360

$Sc = \frac{\mu}{\rho D} = \frac{\nu}{D}$ = 412,440

$Sh^{(2)}$: predicted by Eisenberg mass transfer model : 1,810; $Sh^{(2)} = 0.079\ Re^{0.7}\ Sc^{0.356}$

$k^{(2)}$ cm/s = $\frac{ShD}{d}$ = 1.51 x 10^{-4}; mass transfer factor, $sh^{(2)}/Sc^{0.356}$ = 18.14

t=time (hours)	I_L measured (mA)	CR (mm/y) measured ($CR^{(1)}$)	Cb (x 10^{-6} mols/cm^3)	$k^{(1)} = \frac{CR^{(1)}}{A(C_s - C_b)}$ cm/s x 10^{-4}	$Sh^{(1)} = \frac{k^{(1)}d}{D}$	$I_{L_{CALC}}$ (mA) $= k^{(2)} z F (C_s - C_b) S$	CR_{CALC} (mm/y) ($CR^{(2)}$)	$\frac{CR^{(1)} - CR^{(2)}}{CR^{(2)}}$(%)	$Sh^{(1)}/Sc^{0.356}$
18	1.30	1.54	0.179	3.27	3,922	0.598	0.71	+117	39.30
26	1.05	1.24	0.201	2.67	3,022	0.592	0.70	+77	30.28
42	0.90	1.07	0.230	2.32	2,782	0.584	0.69	+54	27.88
49	0.85	1.01	0.280	2.25	2,698	0.569	0.67	+49	27.04
66	0.65	0.77	0.430	1.85	2,218	0.528	0.62	+28	22.22
71	0.60	0.71	0.430	1.70	2,045	0.528	0.62	+13	20.49
88	0.54	0.64	0.466	1.56	1,870	0.517	0.61	+4	18.74

$k^{(1)}$, $Sh^{(1)}$: mass transfer factor and Sherwood number determined from limiting current measurement

Table D.28

Hydrodynamic Corrosion Data for Mondi B in 98% H_2SO_4 at 27°C (81°F) and 1.62 ms^{-1}

	Alloy MONDI B	Solution 98%H_2SO_4	Velocity (ms^{-1}) (rpm) 1.62 1,000	T° (°K) (°C) 300 27	Exposure Time t (hours) 77	S = 9.74cm^2 d = 3.1cm
μ : viscosity (kg/ms^{-1}) (CP) 19.5 x 10^{-3} 19.5; D : diffusion coefficient (m^2/s) 25.85 x 10^{-12}	ρ : density (kg/m^3) 1,829		C_s surface concentration (x 10^{-6} $mols/cm^3$) 2.288		F=96,500 cb M = 55.8 g/mols Z=2 A=2.24.10^9($mols/cm^2/s/mm/y$)$^{-1}$	

$Re = \frac{dv\rho}{\mu} = \frac{dv}{\nu}$ = 4,725

$Sc = \frac{\mu}{\rho D} = \frac{\nu}{D}$ = 412,440

$Sh^{(2)}$: predicted by Eisenberg mass transfer model : 2,942 $Sh^{(2)} = 0.079\, Re^{0.7}\, Sc^{0.356}$ — $k^{(2)}$ cm/s = $\frac{ShD}{d}$ = 2.453 x 10^{-4} — mass transfer factor, $sh^{(2)}/Sc^{0.356}$ = 29.49

t=time (hours)	I_L measured (mA)	CR (mm/y) measured ($CR^{[1]}$)	Cb (x 10^{-6} $mols/cm^3$)	$k^{(1)} = \frac{CR^{(1)}}{A(C_s - C_b)}$ cm/s x 10^{-4}	$Sh^{(1)} = \frac{k^{(1)}d}{D}$	I_{Lcalc} (mA) $= k^{(2)}\, z\, F\, (C_s - C_b)\, S$	CR_{CALC} (mm/y) ($CR^{[2]}$)	$\frac{CR^{(1)} - CR^{d}}{CR^{(2)}}$(%)	$Sh^{(1)}/Sc^{0.356}$
3	0.65	0.77	0.488	1.909	2,290	0.830	0.98	-22	22.95
12	0.64	0.76	0.502	1.899	2,277	0.823	0.97	-22	28.84
24	0.58	0.69	0.574	1.797	2,155	0.790	0.94	-26	21.60
29	0.63	0.75	0.574	1.950	2,338	0.790	0.94	-20	23.43
48	0.52	0.62	0.717	1.760	2,111	0.724	0.86	-28	21.15
56	0.42	0.49							
75	0.42	0.49	0.753	1.450	1,739	0.707	0.84	-40	17.42
77	0.42	0.49	0.789	1.486	1,782	0.691	0.82	-39	17.86

$k^{(1)}$, $Sh^{(1)}$: mass transfer factor and Sherwood number determined from limiting current measurement

Table D.29

Hydrodynamic Corrosion Data for Mondi B in 98% H_2SO_4 at 27°C (81°F) and 2 ms^{-1}

Alloy	Solution	Velocity (ms^{-1}) (rpm)	T° (°K) (°C)	Exposure Time t (hours)	
MONDI B	98%H_2SO_4	2 1,235	300 27	68	S = 9.74cm^2 d = 3.1cm

μ : viscosity (kg/ms^{-1}) (CP)	D : diffusion coefficient (m^2/s)	ρ : density (kg/m^3)	C_s surface concentration ($\times 10^{-6}$ $mols/cm^3$)	
19.5 x 10^{-3} 19.5	25.85 x 10^{-12}	1,829	2.288	F=96,500 cb M = 55.8 g/mols Z=2 A=2.24.10^9($mols/cm^2/s/mm/y)^{-1}$

$Re = \frac{dv\rho}{\mu} = \frac{dv}{\nu}$ = 5,815

$Sc = \frac{\mu}{\rho D} = \frac{\nu}{D}$ = 412,440

$Sh^{(2)}$: predicted by Eisenberg mass transfer model : 3,403
$Sh^{(2)} = 0.079\ Re^{0.7}\ Sc^{0.356}$

$k^{(2)}$ cm/s = $\frac{ShD}{d}$ = 2.838 x 10^{-4} mass transfer factor, $sh^{(2)}/Sc^{0.356}$ = 34.10

t=time (hours)	I_L measured (mA)	CR (mm/y) measured ($CR^{[1]}$)	Cb (x 10^{-6} $mols/cm^3$)	$k^{(1)} = \frac{CR^{(1)}}{A(C_s - C_b)}$ cm/s x 10^{-4}	$Sh^{(1)} = \frac{k^{(1)}d}{D}$	$I_{L_{CALC}}$ (mA) $= k^{(2)} z F (C_s - C_b) S$	CR_{CALC} (mm/y) ($CR^{[2]}$)	$\frac{CR^{(1)} - CR^{(2)}}{CR^{(2)}}$(%)	$Sh^{(1)}/Sc^{0.356}$
5	0.460	0.54	0.875	1.728	2,072	0.753	0.896	-39	20.76
41	0.400	0.47	0.889	1.518	1,820	0.745	0.886	-46	18.24
53	0.340	0.40	0.860	1.266	1,518	0.760	0.905	-55	15.21
68	0.315	0.37	0.860	1.172	1,405	0.760	0.905	-58	14.08

$k^{(1)}$, $Sh^{(1)}$: mass transfer factor and Sherwood number determined from limiting current measurement

Table D.30

Hydrodynamic Corrosion Data for Ductile Iron in 98% H_2SO_4 at 40°C (104°F) for the Velocity Range Used

Velocity (ms^{-1})	Time (hours)	$I_{L\,MEAS}$ (mA)	k_{MEAS} (x 10^{-4} cm/s)	C_b (x 10^{-6} moles/cm^3)	$I_{L\,CALC}$ (mA)	$x = \frac{I_{L\,MEAS}}{I_{L\,CALC}}$	k_{CALC} (x 10^{-4} cm/s)	$P = \frac{1-x}{k_{CALC}x}$	k film $= \frac{k_{MEAS} \cdot k_{CALC}}{k_{CALC} - k_{MEAS}}$ (x 10^{-4}cm/s)
0.81	2	5.40		0	1.87		2.192		
	7	3.37							
	10	2.90		0.322	1.73				
	23	2.10		0.537	1.65				
	24	2.0							
	47	1.52		0.788	1.54	0.987		60	
	59	1.45	2.056	0.788	1.54	0.941		286	33.11
1.62	1	1.65					3.561		
	2	1.65		0.800	2.43	0.679		1,327	
	12	1.57		0.968	2.38	0.659		1,453	
	16	1.57							
	36	1.45	2.298	1.183	2.24	0.647		1,532	6.48
2	1	1.50		1.200	2.58	0.581	4.120	1,750	
	8	1.55		1.304	2.50	0.620		1,487	
	27	1.45	2.543	1.505	2.34	0.619		1,494	6.64
	29	1.26	1.804	0.824	2.87	0.439		3,101	3.21
	30	1.30							
	36	1.26	1.894	1.000	2.73	0.460		2,850	3.50

Table D.31

Hydrodynamic Corrosion Data for Mondi A in 98% H_2SO_4 at 40°C (104°F) for the Velocity Range Used

Velocity (ms^{-1})	Time (hours)	$I_{L\ MEAS}$ (mA)	k_{MEAS} ($\times 10^{-4}$ cm/s)	C_b ($\times 10^{-6}$ moles/cm^3)	$I_{L\ CALC}$ (mA)	$X = \frac{I_{L\ MEAS}}{I_{L\ CALC}}$	k_{CALC} ($\times 10^{-4}$ cm/s)	$P = \frac{1 - X}{k_{CALC} X}$	$k\ film = \frac{k_{MEAS} \cdot k_{CALC}}{k_{CALC} - k_{MEAS}}$ ($\times 10^{-4}$cm/s)
0.81	1	6.10		0	1.87		2.192		
	2	4.50							
	5	2.84							
	12	2.11		0.357	1.72				
	24	1.60							
	28	1.60		0.428	1.69	0.946		260	
	39	1.47		0.571	1.63	0.901		501	
	50	1.41							
	61	1.38	1.912	0.700	1.58	0.873		663	14.97
1.62	1	1.60		0.714	2.56	0.625	3.561	1,684	
	4	1.60							
	10	1.53							
	16	1.49		1.035	2.34	0.636		1,607	
	24	1.44	2.231	1.106	2.29	0.628		1,663	5.97
	25	1.40		0.750	2.53	0.553		2,270	
2	1	1.46		0.800	2.90	0.503	4.120	2,398	
	2	1.44							
	13	1.4		0.964	2.76	0.507		2,360	
	33	1.23	1.906	1.107	2.65	0.464		2,804	3.55

Table D.32

Hydrodynamic Corrosion Data for Grey Cast Iron in 98% H_2SO_4 at 40°C (104°F) for the Velocity Range Used

Velocity (ms^{-1})	Time (hours)	$I_{L\,MEAS}$ (mA)	k_{MEAS} ($\times 10^{-4}$ cm/s)	C_b ($\times 10^{-6}$ moles/cm^3)	$I_{L\,CALC}$ (mA)	$x = \frac{I_{L\,MEAS}}{I_{L\,CALC}}$	k_{CALC} ($\times 10^{-4}$ cm/s)	$P = \frac{1 - x}{k_{CALC} x}$	$k\ film = \frac{k_{MEAS} \cdot k_{CALC}}{k_{CALC} - k_{MEAS}}$ ($\times 10^{-4}$cm/s)
0.81	1	5.70		0	1.87		2.192		
	5	4.20							
	6	3.50							
	7	3.40		0.150	1.81				
	12	2.90		0.215	1.78				
	24	1.95		0.429	1.69				
	27	1.90							
	37	1.50		0.536	1.65	0.910		451	
	49	1.35	1.914	0.787	1.55	0.870		684	15.08
1.62	1	1.45		0.787	2.50	0.580	3.561	2,033	
	4	1.40							
	12	1.30		0.858	2.46	0.528		2,510	
	24	1.18	1.811	1.073	2.32	0.508		2,720	3.686
	28	1.10		0.572	2.65	0.415		3,958	
	35	1.10		0.679	2.58	0.426		3,784	
2	2	1.09		0.679	2.98	0.365	4.120	4,222	
	12	0.92							
	24	0.60		0.929	2.79	0.215		8,862	
	34	0.73	1.097	1.000	2.73	0.267		6,663	1.495

Table D.33

Hydrodynamic Corrosion Data for Mondi B in 98% H_2SO_4 at 40°C (104°F) for the Velocity Range Used

Velocity (ms^{-1})	Time (hours)	$I_{L\,MEAS}$ (mA)	k_{MEAS} (x 10^{-4} cm/s)	C_b (x 10^{-6} moles/cm^3)	$I_{L\,CALC}$ (mA)	$x = \frac{I_{L\,MEAS}}{I_{L\,CALC}}$	k_{CALC} (x 10^{-4} cm/s)	$P = \frac{1-x}{k_{CALC}x}$	k film $= \frac{k_{MEAS} \cdot k_{CALC}}{k_{CALC} - k_{MEAS}}$ (x 10^{-4}cm/s)
0.81	5	3.90			1.87		2.192		
	12	2.80							
	23	1.90							
	36	1.38		0.585	1.63	0.847		824	
	47	1.35	1.870	0.700	1.58	0.854		780	12.72
	52	1.25		0.350	1.72	0.726		1,722	
	58	1.28		0.393	1.70	0.753		1,496	
1.62	1	1.27		0.400	2.77	0.458	3.561	3,323	
	11	1.38							
	17	1.39							
	24	1.34		0.714	2.56	0.523		2,561	
	36	1.35							
	47	1.28							
	59	1.20	1.768	0.928	2.41	0.498		2,831	3.51
2	1	1.17		0.928	2.79	0.419	4.120	3,366	
	3	1.14							
	10	1.11							
	23	1.06		1.035	2.71	0.391		3,780	
	34	0.98		1.070	2.68	0.365		4,222	
	51	0.82	1.271	1.107	2.65	0.309		5,428	1.84

Table D.34

Measured Corrosion Rates Expressed as $\frac{Sh}{Sc^{0.356}}$ for All Four Cast Irons Tested in 98% H_2SO_4 at 40°C

ALLOY / V (ms^{-1})	DUCTILE IRON		MONDI A		MONDI B		GREY CAST IRON	
	Sh	$\frac{Sh}{Sc^{0.356}}$	Sh	$\frac{Sh}{Sc^{0.356}}$	Sh	$\frac{Sh}{Sc^{0.356}}$	Sh	$\frac{Sh}{Sc^{0.356}}$
0.81	1667.0	21.68	1550.0	20.16	1515.9	19.72	1551.6	20.18
1.62	1863.4	24.24	1809.0	23.53	1433.0	18.64	1468.0	19.09
2.0	2061.0	26.81	1546.0	20.11	1030.0	13.40	889	11.57

Table D.35

Calculated Values for K_{MEAS}, K_{FILM}, and P for the Cast Irons Tested in 98% H_2SO_4 at 27°C

ALLOY / V (ms^{-1})	DUCTILE IRON			MONDI A			MONDI B			GREY CAST IRON		
	K_{MEAS} x 10^{-4} cm/s	K_{FILM} x 10^{-4} cm/s	P s/cm	K_{MEAS} x 10^{-4} cm/s	K_{FILM} x 10^{-4} cm/s	P s/cm	K_{MEAS} x 10^{-4} cm/s	K_{FILM} x 10^{-4} cm/s	P s/cm	K_{MEAS} x 10^{-4} cm/s	K_{FILM} x 10^{-4} cm/s	P s/cm
0.81	1.35	12.74	736	1.377	15.63	575	1.51	—	0	1.00* 1.266	2.96 7.84	3,411 1,260
1.62	1.777 1.712	6.45 5.66	1,587 1,747	2.266	29.72	306	1.486	3.80	2,610	1.199 1.706	2.35 5.60	4,243 1,747
2	1.859 2.386	5.38 14.98	1,815 622	2.191 1.884	9.61 5.60	994 1,735	1.17	2.00	4,866	1.180 1.923	2.02 5.97	4,866 1,658

*Note: The two values reported for each parameter correspond in order to Working electrode 1 and Working electrode 2.